Public Private Partnership for Desertification Control in Inner Mongolia

D1796848

Zhongju Meng • Xiaohong Dang • Yong Gao

Public Private Partnership for Desertification Control in Inner Mongolia

Science Press
Beijing

Springer

Zhongju Meng
Desert Control Science and Engineering
Inner Mongolia Agricultural University
Hohhot, Nei Mongol, China

Xiaohong Dang
Desert Control Science and Engineering
Inner Mongolia Agricultural University
Hohhot, Nei Mongol, China

Yong Gao
Desert Control Science and Engineering
Inner Mongolia Agricultural University
Hohhot, Nei Mongol, China

ISBN 978-981-13-7501-9   ISBN 978-981-13-7499-9   (eBook)
https://doi.org/10.1007/978-981-13-7499-9

This Springer imprint is published by the registered company Springer Nature Singapore Pte Ltd.
The registered company address is: 152 Beach Road, #21-01/04 Gateway East, Singapore 189721, Singapore

Foreword

The achievements are funded by the Science and Technology Major projects of Inner Mongolia (2018058); National Natural Science Foundation of China (51769019).

Judging from the actual situation in our country, preventing land degradation and innovating land management activities to promote sustainable development belong to the category of public works. Public-private partnerships have broad application and development prospects in this field. Inner Mongolia is one of the most degraded provinces in China. After unremitting efforts, China has made some achievements. With the deepening of public participation and ecological construction, private enterprises' investment in land degradation management has made up for the shortage of government investment. Public-private partnerships to combat land degradation were first developed in the region. It has produced ecological, social, and economic benefits. In order to better promote the development of ecological construction and strengthen the control of land degradation, this subject has systematically studied the public-private partnership to prevent and control land degradation in Inner Mongolia Autonomous Region through field research and case analysis. The purpose of this study is to explore public-private partnerships that can help prevent land degradation and expand investments in sustainable land management.

Professor, Department of Desert Control Science and Engineering, Inner Mongolia Agricultural University, Hohhot, China, Email: mengzhongju@126.com   Zhongju Meng

Associate professor, Department of Desert Control Science and Engineering, Inner Mongolia Agricultural University, Hohhot, China, Email: dangxiaohong1986@126.com   Xiaohong Dang

Professor, Department of Desert Control Science and Engineering, Inner Mongolia Agricultural University, Hohhot, China, Email: 13948815709@163.com   Yong Gao

Contents

Chapter 1
Introduction

1.1 Research Background

1.1.1 International Background

After World War II, European and American countries embarked on the road of development that was later called "welfare state," characterized by a large part of the national economy leaning toward nationalization. Before the war, especially in European and American countries before the Great Depression, the liberalized market economy was promoted. After the Great Depression, it turned to the Keynesian path of active government intervention. Welfare system has become the main institutional choice of the left-wing parties, especially in Western Europe and Northern Europe. For example, Britain began to establish a welfare state system from cradle to grave in 1948. Judging from the background at that time, the advantages outweigh the disadvantages. However, in the early 1970s, especially when the oil crisis broke out, the original welfare system continued to suffer. As Mr. Giddens pointed out: "Capital owners have lost a lot of power. The state controls most of the decision-making power and resources previously held by private capital, resulting in the combination of nationalization and the development of welfare system." When the pendulum swings to its peak, it will move in the opposite direction. Under this background, since the conservative leader Margaret Thatcher came to power in 1979, both the conservative party and the labor party cabinet have strongly advocated the reform path of "privatization." Similarly, the Republican Ronald Reagan's revolution in 1980 during Reagan's administration, both of which claimed the rationality of the reform itself with "neoliberalism," the original color of state intervention gradually faded out, and the market economy returned to the major European and American countries as a "panacea" to cure the long-term stagnation of the domestic economy. However, this market-oriented reform is not only different from the pure laissez-faire market economy in the past but also from the "Keynesianism" in the

© Science Press & Springer Nature Singapore Pte Ltd. 2020
Z. Meng et al., *Public Private Partnership for Desertification Control in Inner Mongolia*, https://doi.org/10.1007/978-981-13-7499-9_1

1930s, and as Anthony Giddens said, "beyond the left and right" to take the "third way."

In the early days of the Thatcher government, Margaret Thatcher began to supply the existing state-owned assets, especially the original public goods and services, mainly referring to the roads and toll roads directly produced and supplied by the government before the oil crisis. Public goods, such as telecommunications infrastructure, bridges, tunnels, schools, railways, subways, waterworks, etc., have begun to encourage the private sector to finance, design, build, and operate. From the early 1980s till now, public-private partnership has experienced about 30 years of development. As this kind of franchise system is usually designed for 20 and 35 years in countries such as Europe and the United States, in recent years, this new system has tested the early public-private partnership. From the initial ideal design to the later mixed results, the facts tell us that the government initially assumed the legislative responsibility and regulatory responsibility, while the private sector used the market mechanism and competition mechanism to finance, design, and operate various institutional arrangements for the supply of public goods that are good and bad.

In recent years, some western scholars pointed out that with the expiration of the first batch of PFI projects in Britain, judging from the results of the performance evaluation, the success rate of hundreds of large- and medium-sized public-private partnership schemes involving hundreds of billions of pounds is only about 50%, which is significantly different from the concept of public-private partnership and the original intention of its system design. Especially after 2007, the global economy has been hit hard by the US financial crisis. Both public finance and private capital have been affected, gradually affecting the real economies of various countries. In the past, the public and private sectors used the advantages of sharing policy resources, public resources, capital, technology, human resources, and management techniques. The concept and experience of realizing the best supply of public goods and services have been re-examined by the world. Especially in the process of economic and political system transformation, the majority of developing countries have adopted more "borrowing" attitude. The reform of western public management and governance structure often adopts the reform process of "taking its form and losing its spirit." Therefore, when the innovative governance structure of western countries is alienated, it is even more at a loss.

1.1.2 Domestic Background

China's public-private partnership is basically the same as the reform and opening-up process in 1978. Some people think that China's reform and opening-up policy is basically synchronized with the wave of privatization reform in Europe and the United States dominated by western neoliberal political trends in the 1970s and 1980s. In other word, Deng Xiaoping's "reform and opening up" is that same as Margaret Thatcher's "privatization" reform in Britain. The main purpose of

"Reagan" in the United States is the same. We think this understanding is basically correct, but it ignores the significance of the starting point and ultimate goal of China's market-oriented reform. On the surface, China's reform and opening up is an endogenous change in its economic system. It pursues the transition from a planned economy to a commodity economy and a market economy. This is China's pursuit in the 30-year planned economy. This is a historical change since the "Shang Yang Reform." It is a typical country that the market-oriented reform in the era of welfare economy is essentially opposed to excessive nationalization. In any case, this is a "denationalization" campaign against state-owned enterprises. In essence, China's market-oriented reform is a zero-point exploration of the market-oriented road. We are starting a "revolution" on the basis of poverty. In the period of planned economy, the supply of public goods in China, especially the supply of local public goods and services, was basically completed by local governments at all levels and various state-owned units. This government supply mechanism was later called administrative monopoly supply. However, due to the long-term shortage of various consumer goods after the founding of the People's Republic of China, after the reform and opening up, the introduction of foreign capital and the revitalization of the individual economy, the private economy is mainly to solve the contradiction between the growing material and cultural needs of the masses and backward productivity. In other words, in the early stage of the reform, the supply of public goods has not officially entered the attention of government decision-makers at all levels. At the beginning of the reform, a small number of foreign investment projects used foreign capital's technology, management experience, and capital advantages to develop local public utilities. These projects were only taken as part of "foreign direct investment" by the central government and included in the current account or part of foreign trade surplus. At this stage, the market-oriented reform of public goods and services has not yet reached the stage of transformation of government governance mode, upgrading of management methods and follow-up of comprehensive policies and regulations. Therefore, public-private partnerships are rarely mentioned in the theoretical circle. In the 1980s, BOT project mode of public infrastructure entered the expressway field and was considered as the originator of public-private cooperation in China. However, this model was only one of the sub-models that were too mature in the public-private partnership in the Western world at that time.

In the reality of land degradation control, due to the influence of climate change and human factors, the ecological environment in Northwest China has been deteriorating. Although the government has invested a lot of money, and also issued a number of preferential policies, it has not fundamentally changed the situation of "partial improvement and overall deterioration." The deteriorating ecological environment not only hinders the economic development and the improvement of people's living standards in Northwest China but also seriously affects the economic growth of the region and the security of the whole country. To curb the trend of ecological deterioration and reconstruct the ecological environment in Northwest China has become an arduous task in front of us. At the same time, ecological environment management has become a remarkable technical and economic activity since the twentieth century. Researchers continue to explore ways and means to improve the performance of ecological environment governance from the technical, economic, and even social point of view. Because the economic behavior of the governance subject was different under different institutional provisions, the institutional performance may also be quite different. Therefore, through the innovation of ecological environment management system, the institutional arrangements conducive to ecological restoration and reconstruction are designed to improve the efficiency of the use of ecological construction funds and stimulate people to invest in ecological environment management. It is of great theoretical and practical significance to protect the existing ecological resources.

Ecological construction is a huge project, which has a wide range of governance, difficult to manage, and has the nature of social public welfare. The state should be the main body of investment and increase the intensity of transfer payment. However the national financial resources are limited, effective organization is lacking, and the governance effect is not significant. At the same time, people also realize that the bad ecological environment can not only improve the environment but also produce considerable economic benefits, which is actually one of the main driving forces to attract individual enterprises to engage in ecological construction. Moreover, both individuals and enterprises manage the ecological environment to solve the problem of responsibility and rights, especially the enterprise operation has the advantage of forming ecological industry.

Enterprises can develop tens of thousands of acres of barren mountains and deserts at the same time, and the scale of governance is large and lasts a long time, which can make a breakthrough in the forest and grass industry in northwest China.

The enterprise operation type is stronger than the individual contract type capital; the technology and the marketing ability is more standard than the government impetus-type development pattern. At the same time, competition also makes enterprises strengthen technology research and development, provide more competitive green products to the market, and so on.

Therefore, the ecological environment management country should turn the public welfare management to the benefit management and encourage the social forces to invest in the ecological management.

As far as the actual situation of China is concerned, preventing and controlling land degradation and carrying out innovative sustainable land management activities belong to the category of public utilities. PPP has broad application and development prospects in this field. Inner Mongolia is one of the provinces with the most serious land degradation in China. After unremitting efforts in the early stage, some achievements have been made.

With the participation of the whole people and the increasing intensity of ecological construction, private enterprises have invested in land degradation prevention and control and ecological construction to a certain extent to make up for the lack of government investment. And the public-private partnership to prevent and control land degradation has been initially formed in our region and produced good ecological, social, and economic benefits. In order to better promote the cause of ecological construction in our region and expand the participants in land degradation prevention and control, this topic makes a systematic study on the public-private partnership for land degradation prevention and control in Inner Mongolia Autonomous Region through field research and case analysis. The purpose of this paper is to explore and study a public-private partnership model in line with the local reality in order to effectively expand the scale of investment in land degradation prevention and sustainable land management.

1.2 Summary of Research

PPP involves many stakeholders, such as the public sector, the private sector, and the public, and occurs in complex political, economic, and social environments. To understand how PPP can better promote the success of PPP in practice, many scholars have carried out in-depth research from many angles; these studies involve economics, management, public administration, and other disciplines, the following from these aspects are reviewed.

1.2.1 "PPP" Mode Connotation

1.2.1.1 Basic Overview

The Chinese full name of PPP model is "public-private partnership," which originates from English "public-private partnerships" and is referred to as public-private partnership in Chinese. PPP itself was a very broad concept; coupled with ideological differences, it has been difficult to reach a consensus on the exact meaning of PPP in the world. Norbert Portz, a German scholar, even thinks that there is no meaning in trying to sum up what PPP is or should be, that it does not have a fixed definition, and that it is difficult to examine the origin of this vague English word. The exact meaning of PPP should be determined according to different cases. For the concept of PPP mode, different scholars and institutions have different statements. As defined by the United Nations Institute for Training and Research, PPP covers all forms of institutionalized cooperation among advocates of different social systems, with the aim of addressing certain complex local or regional issues. PPP has two meanings: one is the cooperation between public and private advocates to meet the needs of public goods, and the other is to meet the needs of public goods. The fact that large public projects are carried out in partnerships between the public and private sectors. As defined by the European Commission, PPP refers to a cooperative relationship between the public and private sectors; the purpose of which is to provide public projects or services traditionally provided by the public sector. The Canadian Public-Private Partnership Association (Canadian Council for Public-Private Partnerships) is defined as the PPP model as a risk cooperation between the public and private sectors, which is based on the expertise of various partners and through resources. The proper distribution of risks and benefits to best meet public needs. According to the definition of the PPP national committee of the United States, PPP is a form of provision of public goods between outsourcing and privatization, and combining the characteristics of both makes full use of private resources to design, construct, invest, operate, and maintain public infrastructure and provide services to meet public demand. Kernighan defines the PPP model as a broad relationship, that is, PPP is the relationship between public and private sectors to share each other's rights, work, support, and information in order to share their common goals or interests. Cao Yuanzheng's definition of PPP model in the system of public and private cooperation and its application in China is that the core content of PPP is to complete some public facilities. Public transport and related services project to reach a partnership between public and private institutions, signing contracts to clarify the rights and obligations of both parties to ensure the smooth completion of these projects. The basic characteristics of PPP include the sharing of investment returns, the sharing of investment risks, and the assumption of social responsibility. See Table 1.1.

Ramina S. believes that public-private partnerships are emerging in an innovative form, a close partnership between different partners allied under the project, established by the public and private sectors in the direction of output efficiency and sustainable development, and carry on the successful management. David and Klaus believe that public-private partnerships are aimed at achieving an established goal in which public and private parties act accordingly, provide their own resources, form institutional structures under partnerships and complement each other with public-private advantages, and collaborate to complete products and projects that require specific expertise and a public-private partnership works for a team. Shafiul believes that public-private partnerships are the private sector responsible for services, the public sector to regulate and safeguard the public interest. Through public-private partnerships, the private sector has the advantages of investment and financing, technology, management efficiency, and entrepreneurship, and the public sector as a policy orientation has the advantages of social responsibility, environmental protection, local identity, and concern for employment. The public and private sectors take on the social role in the form of complementarity, give full play to their respective advantages, and form in the form of coalitions. To achieve sustainable and effective public-private partnerships, the legal system and regulation are

Table 1.1 The comparison between the PPP mode public sector and the private sector

Public sector	Private sector	Partnerships
Policy – demand – decree	Enterprise management	Risk allocation and transfer
Political commitment	Project management and experience	Integrated goals
Social and cultural acceptance	Technological innovation	Delivery guarantee
Government reputation	Own funding requirements	An agreement to obtain a financing loan
Statutory structure	Organization support	Spirit of teamwork

necessary to enhance partnerships with possible subsidies. Two forms of cooperation in public-private partnerships, namely, the formation of joint ventures and the establishment of contractual relationships, are mainly used in technical and financing cooperation. It was also argued that the public sector refers to the sum of the government and its subordinate departments, implying three sectors, namely, the public sector, the public utility sector, and the public enterprise sector, and the public sector in our country has the characteristics of centralization and decentralization. The private sector includes private businesses, individual investors, and foreign multinational corporations. The private sector participates in the investment, financing, and management of public projects through the purchase of equity, municipal bonds, franchises, and the construction and operation of public projects.

In general, the PPP model refers to the long-term partnership established by the public and private sectors to provide public services. In PPP, the public and the private sectors sign agreements to provide public services, give full play to their respective advantages, share risks, and share benefits. PPP can be divided into broad sense and narrow sense. PPP in a broad sense comes in many forms and can include all forms of public service delivery between the full government and the private sector (Fig. 1.1), including service outsourcing, joint ventures, franchising and government subsidies for private development projects, etc. In this continuum, at the far left is the pattern provided entirely by the government and at the far right is the pattern provided entirely by the private sector. The differences between the different forms in the right half of the continuum are relatively small and vary from case to case.

In a broad sense, PPP can also be divided into three categories: outsourcing, franchising, and privatization. Outsourcing PPP projects are usually invested by government departments and owned by government departments, and one or more of the functions of the project are contracted by the private sector, such as engineering design and construction. Or by the government commissioned to manage and maintain the project facilities, the private sector is less motivated, the risk is relatively small, and the private sector income mainly comes from the government department to pay.

Franchise category PPP project refers to the private sector to provide some or all of the investment and through a certain cooperation mechanism to share risks and benefits with government departments. The private sector has a higher participation in the project and is responsible for the construction, operation, and management of

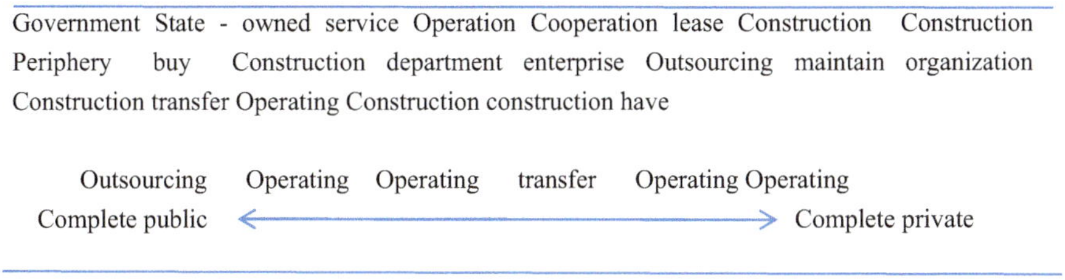

Government State - owned service Operation Cooperation lease Construction   Construction
Periphery   buy   Construction department enterprise Outsourcing maintain organization
Construction transfer Operating Construction construction have

Outsourcing   Operating   Operating   transfer   Operating Operating
Complete public ⟵─────────────────────────────⟶ Complete private

Fig. 1.1 Public-private partnership type continuum

the whole project, while the government department gives the private sector a certain amount of compensation or charges it a certain period of concession fees according to the actual income of the project. The ownership of the whole project will return to the government after the expiration of the franchise. In this model, the enthusiasm of the private sector is relatively high, but it is very important for government departments to find a balance between the public welfare of PPP projects and the profitability of the private sector, which puts forward higher requirements for the management ability of government departments. Under this mode, the management ability of government departments even restricts the success or failure of the project to a certain extent. If an effective supervision mechanism can be established, the model can give full play to the respective advantages of the government and the private sector. The typical mode of franchising PPP is bulid-operate-transfer (BOT) model.

The privatization of PPP projects which refers to all project investments is the responsibility of the private sector; under the supervision of the government department, the private sector maintains the normal operation of the project by charging users and obtains the corresponding return while recovering the investment. The ownership of the project in this way is owned by the private sector, so the private sector also bears the greatest risk in this approach. However, when the management level of government departments is good, this method is a special incentive for the private sector, which can ensure the quality of public goods and the durability of their time. The specific forms of PPP are bulid-operate-transfer (BOT), build-transfer(BT), build-transfer-operation(BTO), build-own-operate (BOO), bulid-own-operate-transfer (BOOT), buy-build-operate (BBO), lease-build-operate (LBO), expansion to manage the overall project and transfer (wrap-around addition), service agreement (service contract), operation and maintenance agreement (operate and maintenance contract), and so on. However, the various forms of PPP are only means, and improving the efficiency of the supply of public goods is the key. In a narrow sense, PPP can be understood as the general term of a series of project financing models, which refers to the institutional arrangement in which the public sector and the private sector participate in the production and provision of goods and services, and is a way of project financing, contract contracting and franchising. Subsidies, etc., meet this definition.

The public sector and the private sector work together to provide public services that can achieve their respective goals. The desire of the private sector to participate in the provision of public services in cooperation with the public sector is mainly motivated by the pursuit of profits. The public service industry often has some kind of natural monopoly nature, its competitive pressure and investment risk are relatively small, its income is stable, it can obtain long-term profit, and it has the attraction to the private sector.

The public sector can benefit from private sector funds, technology, management advantages, and so on, mainly in the following areas. First, it can make up for the shortage of public funds. It is the public sector's responsibility to meet the public's demand for a growing number and quality of public services, but the public sector is limited in funding and has insufficient supply capacity, creating a contradiction between the supply and demand of public services.

At present, almost all countries in the world are facing the problem of urgent need to build, maintain, and upgrade infrastructure, but public funds are insufficient. Through private sector investment and financing, the financial pressure on the public sector can be alleviated. To meet the public demand for public services as soon as possible, PPP has become a new choice for the provision of public services. Second, it can improve the efficiency of public services and better realize the value of funds. In the situation of PPP, the distribution of responsibilities and benefits between the public sector and the private sector is relatively clearly defined in the contract, and the private sector has sufficient incentives to reduce the construction and operating costs of public service facilities. The private sector uses its own technical and management advantages to improve efficiency, reduce costs, and make profits, as well as public sector capital expenditure, through the formulation of the best plans, economies of scale and continuous innovation, Give the public access to cheaper services. At the same time, in order to enhance the reputation and long-term competitiveness of the private sector under the framework of PPP, the private sector has the inherent motivation to continuously improve the level of public services, and the quality of public services can also be continuously improved. Private sector's expertise, efficient management, and industry experience provide objective conditions for improving the quality of service.

In addition, the risks in the provision of public services can be reasonably shared. The process of public service is full of risks, including technical risk, construction risk, operational risk, income risk, financial risk, political risk, environmental risk, default risk, and force majeure risk. The private sector can use its own experience to better manage certain risks in business operations and service delivery and reduce the risks borne by the public sector. Therefore, various service delivery methods can also be arranged according to the degree of private sector participation and the amount of risk taken.

The advantages of the PPP model are as follows: first, the public and private sectors participate in the demonstration at the initial stage, and they can determine which public utilities can carry out project financing as soon as possible. And it can better solve the risk allocation throughout the life cycle of the project in the initial stage of the project, because the government shares part of the risk, which makes the risk allocation more reasonable and reduces the risk of contractors and investors, thus reducing the difficulty of project financing. Second, the PPP financing model can enable the private enterprises to be involved in the investment of public utilities and public projects to participate in the early stage of the project, which is conducive to the use of advanced technology and management experience of nonpublic capital. Third, under the PPP model, private capital and public capital participate in the construction and operation of public utilities together, can form a mutually beneficial long-term goal, and can better provide services for the society and the public. Fourth, through the PPP financing model, the participants in public utility projects can be reintegrated, and the strategic alliance can be combined to play a key role in coordinating the different interests and objectives of the parties. The fifth is that the

PPP financing model makes it possible for private enterprises with the intention to participate in the public utility project to come into contact with the government of the local government or the relevant institutions as soon as possible, save the bidding cost, and save the preparation time, so as to reduce the final bid price. Compared with the BOT model, the PPP model has a certain control over the government. It is of great significance to implement the PPP model when China's market economy system is not perfect enough.

PPP model is a kind of cooperative relationship which is guaranteed by legal system, establishes equality and mutual benefit, and can effectively prevent the damage of improper behavior of cooperative parties to the construction and operation of the project. In the PPP model, cooperation will be achieved through legally effective agreements or contract documents signed between enterprises, investors, and the government, which will form a constraint relative to all parties, no matter who fails to perform their contractual obligations. All must bear the prescribed responsibility; these legal systems in turn play a safeguard role in the effective implementation of the agreement or contract.

In a word, the PPP model also has shortcomings: (1) in order to maximize profits, private companies may damage the public interest; (2) the withdrawal of the government from some industries may lead to oligopoly; (3) there is a high cost of contract management and supervision; (4) if government investors collude with private companies, it is easy to lose state-owned assets; (5) due to the defects of evaluation methods, it is difficult to obtain fair prices for the sale of assets; and (6) efficiency and fairness are hard to balance. When the government monopolizes management, due to the absence of asset owners, it leads to inefficient management regardless of cost. Private companies can improve efficiency, but high profits encroach on consumers' interests, which makes the distribution of social wealth unfair.

1.2.1.2 The Background of the PPP Model

Since the 1980s, a new management mode aiming at the defects of traditional public administration mode has swept across the world. It is usually called "new public management" or "managerialism." Its main principles are as follows: management rather than administration, marketization rather than bureaucracy, competition rather than monopoly, result rather than process, etc. Its core contents mainly include improving efficiency, marketization, service orientation and responsibility for policy effects, etc. The theoretical basis of the new public management comes mainly from two aspects, that is, the economic theory and the private sector management theory, but the new public management has also been criticized, such as ignoring the particularity of the public sector, ignoring the policy implementation and government responsibility, and so on; thus the new public service theory has emerged. The main viewpoints of the new public service theory put forward by Denhardt et al. include serving citizens rather than customers, public interest is the goal rather than the by-product, paying more attention to citizenship than entrepreneurship, serving rather than steering, and paying more attention to people rather than productivity. However, the new public service theory is not a theory in a strict sense but a viewpoint or hypothesis, lacking rigorous argumentation and profound theoretical foundation. In short, in the field of public service, the elements of the three modes of public administration, new public management, and new public service coexist. Public administration emphasizes fairness, ethics, and responsibility, new public management emphasizes competition and response to consumers, while new public service pays more attention to citizen participation.

In the field of real government management and public service, there has been a worldwide movement from rights to contracts. The government has been suggested to streamline and deregulate from command-controlled operation to negotiation-driven operation, emphasizing the incentive-oriented process and performance evaluation. From rights-based governance to contract-based governance, not only products and services but also traditional authority-based activities such as mandatory regulation are increasingly completed through agreements. The general trend is that the government is increasingly moving from the use of authoritative mechanisms to consultative governance, including negotiations with regulated enterprises, service contracts with profit-making or nonprofit organizations, cross authority arrangements with other government agencies, service agreements with citizens and performance agreements among members of government organizations, etc. Britain is an early country to promote PPP model. Since the 1990s, the British government has actively reformed its mode of providing public services. It has made full use of the financial, design, and management support and experience of commercial service companies or institutions in various private sectors to provide services for the public sector, forming a complete diversified investment model. In the early 1990s, the British government launched the PFI (private finance initiative), announcing the birth of modern PPP. At that time, British Prime Minister Margaret Thatcher was vigorously promoting privatization. In order to solve the contradiction between the rapid growth of capital demand for infrastructure and the tight public finance, she began to implement PFI plan in the infrastructure field: the government gradually withdrew from many public infrastructure fields and at the same time formulated preferential policies to encourage private capital to enter. Introducing market competition into infrastructure projects, actively attracting private capital to participate in the construction of operating infrastructure, and operating in a market-oriented way can not only effectively reduce the government's expenditure pressure in infrastructure construction but also improve the service level and efficiency of infrastructure. PPP mode is generated under such background and demand.

The British government has vigorously promoted the private finance initiative (PFI) system since 1992. Since the Labor Party came into power in 1997, it has been incorporated into a new concept, public-private partnerships (PPP), and PPP has been interpreted including three aspects: (a) complete or partial privatization, (b) the contracted projects financed by private initiative and taking risks are still called PFI, and (c) providing public services jointly with private enterprises. The British government, through the implementation of PPP/PFI, guides private enterprises to participate in the investment of various infrastructures and, through policies, stipulates that public construction projects should be given priority to public-private cooperation before purchasing, so as to solve the financial difficulties faced by the government.

After the 1990s, this trend has been strengthened in both conservative and moderate governments and in both developed and developing countries. The common point is that the exercise of government power through government agencies has turned to governance through contracts, encouraging market forces to step into areas of direct government action in the past. Public structures act as middlemen rather than direct service providers, strengthening public-private cooperation and replacing coercion and arbitrariness with negotiations and other dispute resolution methods.

1.2.1.3 The Proposed PPP Model

First of all, in theory, although many theories such as transaction cost, principal-agent, and game theory have been applied in PPP analysis, there is still a lack of an appropriate PPP analysis framework, and there is not much knowledge about the factors that promote effective PPP governance and affect PPP success. Further systematic analysis and research are needed.

Secondly, in the aspect of experience, we have examined PPP performance worldwide and found that the conclusions about PPP performance are inconsistent, mixed, and even contradictory, which requires in-depth study of various factors affecting PPP performance.

In addition, developing countries and countries in transition have encountered setbacks and challenges in the process of developing PPP. Estache et al. said in the report on infrastructure operation performance in developing countries and countries in transition that most relevant transnational studies have not found PPP to have significant advantages in efficiency over direct government provision. Since the mid-1980s, nearly 50% of the franchising agreements in Latin America have been renegotiated (Estache 2006). From the perspective of investment in infrastructure, according to Noel's report, the annual investment of the private sector in developing countries peaked in 1997, and the annual investment in transition countries in Europe and central Asia peaked in 2000, with a downward trend thereafter. Kerf's further research found that private sector investment in infrastructure projects in developing countries decreased sharply after the Asian financial crisis in 1997 and showed a downward trend in the following years. Although it began to increase after 2004, it paid more attention to risk reduction. In order to take full advantage of the renewed interest in investment by the private sector, the government needs to establish a reasonable risk allocation mechanism.

1.2.1.4   The Widespread Use of PPP Model

In the 1990s, Britain took the lead in proposing the concept of PPP and actively privatized public services. Later, PPP was gradually widely used in developed countries such as the United States, Canada, France, Germany, Australia, New Zealand, and Japan. In recent years, international organizations such as the United Nations, the World Bank, the European Union, and the Asian Development Bank have also vigorously promoted the concept and experience of PPP in the world. Many developing countries such as India, China, and Brazil and some African countries have also started to learn and practice PPP.

The application scope of PPP has also expanded from traditional public utilities such as roads, urban rail transit, water supply, gas, sewage treatment, and garbage collection to the construction and operation of prisons, schools, hospitals, and sports facilities and even has been applied in the fields of national defense and space flight. Table 1.2 shows the application of PPP in the European continent.

In continental Europe, Britain was the first to put forward the concept of public-private partnership. PFI (private finance initiative) is the most widely used method in practice. Although there are still some disputes, such as the government's responsibility and supervision and the evaluation criteria for PFI projects, PFI has developed rapidly and achieved good results in general. The British Ministry of Finance has evaluated 451 PFI projects that have been completed and entered the operation period and found that 88% of the projects were completed on time or ahead of schedule and the project expenditure did not exceed the public sector budget. Previous studies showed that 70% of non-PFI projects were postponed, 73% of non-PFI project expenditure exceeded the budget, and more than 75% of the respondents believed that PFI projects could meet or exceed the expectations. Focusing on projects with an investment of less than 20 million pounds, the results show that the project construction and operation performance are good, with more than 80% completed on time or ahead of schedule and more than 90% reaching public expectations. According to statistics from the UK Treasury, as of April 2007, the United Kingdom had signed 628 PFI projects with a total investment of 58.56 billion pounds   [http://www.hm-treasury.gov.uk/documents/public_private_partnerships/PPP_PFI_stats.cfm]. PPP is also widely used in the United States. Since the 1990s, PPP in the United States has also been widely used (Table 1.3).

Since the US federal and local governments must provide public services to the growing population without raising taxes, and private sector participation can

Table 1.2   Development of PPP on the European continent

Involved departments	Education, police, defense, port, hospital, airport, rail, rail
	Cars, roads, waste disposal, prisons, bridges, water supply, others
Involved country	Portugal, Sweden, France, Germany, the Netherlands, Belgium, Switzerland, Norway
	Viagra, Finland, Greece, Russia, Slovenia, Spain, Italy, Bulgaria, Hungary

Table 1.3 Low-income countries from 1990 to 2006 private sector

Participation in infrastructure projects (regional ranking)			
Ranked by region based on the number of projects		Area ranking based on investment	
Area	Number of items	Area	Investment (US$million)
Latin America and the Caribbean	1205	Latin America and the Caribbean	438,528
East Asia and the Pacific	1099	East Asia and the Pacific	253,631
Europe and Central Asia	746	Europe and Central Asia	206,813
South Asia	348	South Asia	96,044
Sub-Saharan Africa	332	Sub-Saharan Africa	52,993
Middle East and North Africa	111	Middle East and North Africa	50,576

Source: http://ppi.wordbank.org/explore/ppi_exploreRankings.aspx

improve the quality of public services and reduce growth, the number of PPP projects has increased rapidly. At present, PPP has been extended to almost all public sectors in the United States, such as schools, hospitals, prisons, transportation, garbage disposal, national defense, and aerospace.

1.2.2   Characteristics and Functions of Public-Private Partnerships

1.2.2.1   The Characteristics of Public-Private Partnerships

1.2.2.1.1   Project Objectives Were Consistent

Partnership is the main feature of PPP mode. All successful PPP projects are built on partnership. It can be said that partnership is the most important issue in PPP, without which there is no PPP. The government purchases goods and services, grants authorization, and levies taxes and fines, and the handling of these matters does not necessarily indicate the true existence and continuation of the partnership. For example, even if government departments order sandwiches for lunch from the same restaurant every day, this cannot form a partnership. Compared with other relationships, the partnership between the private sector and the government public sector in PPP is characterized by the consistency of project objectives. The core problem of the cooperation and partnership between the public sector and the private sector is that there is a common goal: to provide the most products or services with the least resources for a specific project. The private sector pursues its own interests with this goal, while the public sector pursues public welfare and interests with this goal. The formation of a partnership should be carried out on the basis of the same project objectives, but this is not enough. In order to maintain the long-term development of this partnership, it is also necessary for partners to consider each other's problems and have two other remarkable characteristics: benefit sharing and risk sharing.

1.2.2.1.2 Benefit Sharing

Benefit sharing is the second characteristic of PPP mode. It should be clearly pointed out that the benefit sharing between the public sector and the private sector in PPP mode is not profit sharing. Any PPP project is a public welfare project and does not aim at maximizing profits. If both sides want to share the profits, it is actually an easy thing. As long as the price is allowed to increase, the profits can be greatly increased. However, doing so will inevitably bring public dissatisfaction and may eventually lead to social chaos.

1.2.2.1.3 Risk Sharing

The third characteristic of PPP mode is risk sharing. Partnership means not only benefit sharing but also risk sharing. Risk sharing is another foundation of partnership besides benefit sharing. Without risk sharing, it is impossible to form such a partnership. No one likes risks whether it is a market economy or a planned economy, a private sector or a public sector, and an individual or an enterprise. In PPP mode, the characteristic that the public sector and the private sector share risks reasonably is a significant sign that it is different from other transaction forms between the public sector and the private sector. For example, the reason why the government procurement process cannot be called a public-private partnership is that both parties let themselves take risks as little as possible in the process. However, in the public-private partnership, the public sector bears as much of the risks associated with its advantages as possible, while the risks borne by the other side are as small as possible. A typical example is that in the construction of tunnels, bridges, and trunk roads, if the private sector fails to achieve the basic expected income due to insufficient traffic flow, the public sector can subsidize its cash flow, which can effectively control the business risks caused by insufficient traffic flow in the private sector under the framework of "sharing." At the same time, the private sector will actually assume more or even all of the specific management responsibilities according to its comparative advantages, and this area, for the public sector, is precisely the area where management's "moral hazard" is easy to occur, thus avoiding this risk. If each kind of risk can be borne by the partner who is best at dealing with the risk, then the cost of the whole infrastructure construction project can be minimized. In PPP management mode, more consideration is given to minimizing the risks of both parties. Facts have proved that the management mode of pursuing the risk minimization of the whole project can resolve the risks in the field of quasi-public goods better than the public and private parties pursuing the risk minimization, respectively. Therefore, the mechanism effect of "one plus one is greater than two" brought by PPP needs to be understood from the level of management mode innovation.

1.2.2.2 The Functions of Public-Private Partnerships

PPP mode is a new management mode, which not only has the general functions of management but also has the functions that other management modes do not have. For example, it has the functions of financing, utilizing new technologies, and innovating mechanisms.

1.2.2.2.1 General Functions

The general functions of PPP model mainly include planning, organization, leadership, and control. Among them, the planning function includes defining organizational goals, developing a global strategy to achieve these goals, and developing a comprehensive hierarchical planning system to integrate and coordinate various activities. Planning is a coordinated plan and process, which points out the direction for managers and non-managers. When all relevant personnel understand the organization's goals and what contribution they must make to achieve them, they can begin to coordinate their activities, cooperate with each other, and form a team. If there is no plan, there will be many detours, thus making the process of achieving the goal ineffective. Plans can be divided into formal plans and informal plans.

In PPP management, the formal plan is jointly formulated by the public and private sectors and approved by contract. Through the plan, we can clearly see what are the common goals of the public and private sectors and what are the respective goals of the public and private sectors. In the process of public-private cooperation, each period also has specific goals. Organizational functions generally consist of organizational structure, organization and position, human resources management, management of change and innovation, etc. Organizational structure describes the framework system of an organization.

In the PPP process, sometimes new organizations will be set up for special projects. Some projects will not set up new organizations based on the original organizations. In the newly established organization, there will generally be both public and private sector personnel, and the corresponding management positions will be arranged according to the contract requirements. The leadership function includes two connotations, one is leadership and the other is leadership. In this respect, PPP also has characteristics different from the general management mode. For example, Shanghai Pudong Waterworks adopts PPP management mode, and the newly established France Vivendi Group holds 50% of the shares in the water company. Its chairman and general manager are in power by rotation between China and France (when Chinese personnel are the chairman, France is the general manager; when France is the chairman of the board, China is the general manager). This reflects the special role and innovative form of leadership function in PPP management mode. Control can be defined as the process of monitoring activities to ensure that they are carried out as planned and to correct various deviations. The more perfect the control system is, the easier it is for managers to achieve their goals. The control process can generally be divided into three steps: first, to measure the actual performance; second, to compare the actual performance with the standard; and, third, to take management actions to correct the deviation. In the process of PPP management, the control function is more obvious. Both the public and private sectors measure actual performance at all times. The private sector measures the actual income (return on investment), while the public sector measures the response of the public. The actual performance of the private sector is compared with other previous projects. The public sector will compare the actual performance with that before the cooperation. When there is a problem in project management, the public and private departments should take corresponding management actions to correct the deviation.

1.2.2.2.2   Special Functions

As a new management mode, PPP is not only the general function of a management but also the function of financing and utilizing new technologies and mechanism innovation, which are not available in general management. Financing function is people's earliest understanding of PPP. Until now, quite a lot of people think PPP is a financing mode. At the beginning of PPP's rise, its main purpose was to finance infrastructure, and its concrete form was mostly the financing of highway construction and railway construction. When the government builds infrastructure such as roads and railways, it often allows the private sector to invest due to insufficient funds, and the private sector recoups its investment through fees. It is this financing function that makes people have great interest and enthusiasm in PPP, and then this PPP financing function is continuously applied to all aspects of infrastructure construction, for example, tap water supply, sewage treatment, tunnel construction, public health and medical treatment, basic education, etc. The public sector of the government provides public goods and services to the society through private capital in different fields, which can make up for the shortage of funds in the process of government providing public goods and services to the society. Construction-operation-transfer mode (BOT) is the most obvious financing function among many PPP management modes. The government public sector usually lets the private sector use its own funds to build infrastructure and then lets the private sector operate and obtain profits from it, which are then transferred to the government department after a certain period of time. The government may not need to invest a cent in this process, but it will provide the society with the infrastructure and services that it should have provided itself. At the same time, after a certain period of time, it will also have the infrastructure. The use of new technologies has two meanings: one is production technology, and the second is the technology of management methods. The reason why it is said that the use of new technologies is a function of PPP management mode is that while providing financing for the public sector, PPP management mode also brings new production technologies and management technologies to the public sector, thus greatly improving the efficiency and level of public goods and services, thus, on the basis of not increasing the public tax burden, relying on the "user payment" mechanism, to meet the needs of the public to the greatest extent with the hands of the private sector. The function of mechanism innovation is mainly to promote mechanism transformation, system innovation, and improvement of resource allocation efficiency in economic and social life.

The essence of PPP is to realize public-private cooperation. The result of cooperation is to form a new management system and operation mechanism superior to the separate role of planning and market. Plans tend to focus more on average, thus losing efficiency, while markets tend to focus more on efficiency, thus losing average. PPP management mode focuses on the organic combination of fairness and efficiency, realizing fairness in social development with as little loss of efficiency as possible, and improving the use efficiency and comprehensive efficiency of economic resources, especially public sector resources, with as little loss of fairness as possible. The function of mechanism innovation highlights the latecomer advantage of PPP management mode and opens up the space for it to play its potential. It can effectively avoid the inefficient detour taken by predecessors to provide public goods or services by planning alone. At the same time, it can overcome the shortage of public investment incentive mechanism and the indifference of the private sector, which are easy to occur under the market economy, and provide an innovative mechanism with obvious "latecomer advantage" characteristics for the provision of public goods and services, the construction of public infrastructure, and the supported social "sound and rapid" development.

1.2.3 Services of Public-Private Partnerships

In public service projects, public services originally provided by government departments directly to the public or society are transferred to the private sector under the PPP mechanism, i.e., the privatization of public services. The original responsibilities, risks, and service returns of the public sector are transferred to the private sector, resulting in different cooperation modes according to the scope of public-private cooperation and the degree of risk sharing. There are two ways to divide PPP project services. One is divided into hard and soft services. Hard services are maintenance and protection directly related to assets. These assets are directly used to provide public services. Their availability affects the revenue mechanism of public services and is an indispensable part of PPP projects. Soft services are the support services for the project, such as cleaning, health maintenance, inquiry, and logistics support, which are not directly related to the available assets of the project. The second is divided into core services and support services. Core service refers to the part where the performance or quality of service provided by application facilities is directly related to economic returns and is the key element of its core competitiveness, such as school teaching and hospital diagnosis and treatment. Support service is the part that provides support and guarantee for core service, which is visually the manual labor service part, similar to the soft service part.

In theory, PPP mechanism should not have obstacles in implementing PPP mode, whether it is hard service or soft service and core service or support service. However, due to the public welfare of public utilities, countries implementing PPP mode may only introduce PPP mode in a limited or partial scope in specific industries according to the specific characteristics of their public utilities. For example, the PPP project in British hospitals only includes soft services such as property management, medical equipment, and logistics services, but does not include core medical services.

1.2.4 Introduction and Implementation of Public-Private Partnerships

1.2.4.1 The Introduction of Public-Private Partnerships

At the same time of PPP mode, it means to introduce relevant systems at the same time, which may require in-depth innovation and reform of public utility-related systems. In order to ensure the effective introduction and implementation of PPP projects, the following five principles should be followed in the operation process:

1. Marketization principle: the selection of partners shall be determined through open, fair, and impartial bidding, so as to ensure the fairness of project implementation procedures.
2. Win-win principle: not only to make the partners profitable but also to strive for the lowest cost of government financing.
3. The principle of combining the introduction of capital with the introduction of intelligence (advanced technology) and systems (advanced management concepts and systems): to examine whether the partners have financial strength and have certain influence in the urban public utilities, whether the brand is prominent, and whether the management is advanced, so as to promote the reform and innovation of PPP project management system and mechanism while attracting capital.
4. The principle of smooth transition: consider ahead of time and properly handle the resettlement of existing personnel to prevent the occurrence of internal unstable factors caused by inadequate propaganda or improper handling.
5. The principle of ensuring government control: the partners must abide by relevant industry regulations, promise to obey the government's industry management, and assume corresponding responsibilities and obligations in social security and public welfare services.

1.2.4.2   The Implementation Process of Public-Private Partnerships

The implementation of PPP projects is relatively complicated, and the specific processes vary from country to country and from project to project. For example, the structure of PPP projects in hospitals in Britain includes the central government, local health administrative departments, senior lenders, project companies and design and construction, property management, and service companies (see Fig. 1.2). The links and processes involved in the specific implementation are even more complicated. However, regardless of the size and category of the project, its commonality is shown as follows: the project initiator plays an intermediary or catalytic role (catalyst). After forming the concept of the project, it seeks partners, establishes a task force, determines a cooperation framework, and establishes the project and its technology, management, and evaluation mechanisms. The framework and stages of its implementation are roughly the same, which can be divided into four stages in principle: The first stage is the formation stage of the plan concept, which includes setting targets, evaluating market potential, and evaluating the company's capabilities. The second stage is the planning and perfection period, which includes partner selection, feasibility study, confirmation of cooperation and signing of MOU (establishment of working group, clear objectives, roles, responsibilities and main partner tasks), preparation of market plan, baseline investigation, validation of market strategy, etc. The third stage is the implementation period. The fourth stage is the evaluation period, which evaluates and summarizes the implementation effect. The implementation process listed by the UK OGC (Office of Government Commerce) includes 14 steps from establishing business needs to contract management.

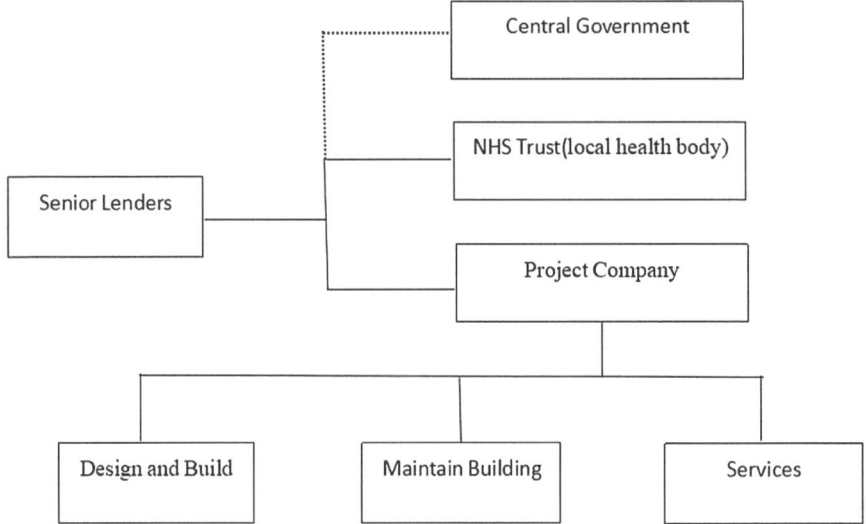

Fig. 1.2 British PPP brief structure

1.2.5 The Importance of Public-Private Partnerships and the Necessity of Research

Traditional theories hold that public infrastructure is a typical market failure area due to its characteristics of regional monopoly and public welfare. It is impossible to play the role of market mechanism. It advocates that within a certain region, a few state-owned enterprises monopolize the operation, resulting in a management system of government-enterprise integration and government monopoly. Public goods operating enterprises have not formed a market-oriented system, have no decision-making autonomy, and have no incentive to reduce costs and improve efficiency. The operating losses of enterprises are subsidized by finance and do not bear operating risks, resulting in low operating efficiency of public utilities. In recent years, the state has issued a series of policies to encourage nonpublic sectors of the economy to enter the public projects and infrastructure areas, gradually improving the franchise system for public projects and infrastructure projects and encouraging nonpublic enterprises to participate in the reform of the property right system and management mode of municipal public enterprises and institutions. This shows that the state's policy to encourage nonpublic enterprises to enter the construction and operation of public projects is increasingly clear, and the policy measures are further specific. At present, various places have carried out market-oriented reforms in such profitable industries as tap water supply, sewage treatment, garbage disposal, and expressways. However, people's understanding of the mechanism of public-private cooperation in public projects is not clear enough, and the design of the cooperation mechanism, risk sharing, and cooperative risk response methods and the establishment of public service price and service supervision mechanism under the public-private partnership are not standardized enough. The marketization reform process of public projects and infrastructure is not standardized, and the management mechanism is not perfect enough, which will affect the effect of marketization and public-private cooperation investment. Therefore, it is necessary to perfect the theoretical basis of public-private cooperation, design the cooperation mechanism between the government and enterprises under the public-private partnership, establish a scientific management framework, ensure the operability of management measures, and promote the success of public-private cooperation.

PPP mode is a public goods supply mode that can make the government, enterprises, and society win-win in many ways. It is an institutional change to transform government functions, integrate social resources, and improve the level of public services. It is also an objective requirement to establish a modern financial system, optimize the allocation of resources, and promote social fairness.

To better play the decisive role of the market in the allocation of resources, the public sector of the government has introduced social capital through PPP to provide public goods and services to the society, forming a mixed ownership economy and realizing a new breakthrough in the property rights of the modern market system. This clear and diversified subject formed by joint-stock system based on joint cooperation can stimulate the decisive role of the market in the allocation of resources to the maximum extent. While effectively alleviating the shortage of government investment, the new management and operation system superior to the separate role of planning and market is formed through the introduction of production management technology formed by private investors based on market factors.

Effectively reducing the pressure on financial expenditure in the next few decades, promoting new urbanization requires a large amount of investment. Judging from the actual situation, infrastructure and public utility investments mostly rely on financial arrangements and government debts. The drawbacks of the investment and financing system, such as gradually widening investment gap, low fund efficiency, and low rate of return on investment, have shown that PPP mode can attract a large amount of private capital and social funds and form a diversified and sustainable fund investment mechanism. It can effectively reduce the expenditure pressure on governments at all levels.

Accelerating the transformation of government functions PPP can organically combine the government's strategic planning, market supervision, management efficiency of public services and social capital, and technological innovation. It is helpful to clarify the boundary between the government and the market, enhance the government's legal awareness and market awareness, and comprehensively improve the level of public management and service.

Promote the establishment of a modern financial system. Finance is the foundation and important pillar of national governance. PPP mode emphasizes the role of market mechanism, the government and social capital do their best, and the deep participation of social capital. It meets the basic requirements of modern financial system to optimize the allocation of resources and promote social equity. Secondly, the long investment return period of PPP projects requires the government to gradually shift from the previous single-year budget revenue and expenditure management to the medium- and long-term financial planning and "asset and liability management," which is conducive to improving the planning and sustainability of finance, preventing and resolving medium- and long-term financial risks, and further improving the infrastructure construction and operation level. PPP mode means that the original monopoly barriers will be removed in the provision of public goods and services, and market competition mechanism will be introduced. Effective competition will improve the efficiency of production and supply of public goods and services. At the same time, private investors, driven by interests and the need to recover investment, are bound to give full play to their advantages in management, improve the level of infrastructure construction and service, and further reduce construction and operating costs.

Private capital enters the field of public investment, but private ownership cannot completely replace public ownership to avoid the formation of new monopolies by private sector management. In order to ensure the higher service quality of public goods and maintain a stable operating environment, the government needs to assume certain social responsibilities, including investment decisions, fund raising, and operation supervision, which shows that the establishment of public-private partnerships for public projects is of great significance.

1. The implementation of PPP mode can meet the needs of investment growth and improvement of investment efficiency. Our country is in the early stage of rapid economic development, with huge investment demand, and financial capacity cannot meet the requirements of infrastructure construction and industrial environmental pollution control. The basic situation of the local credit market is that bank loans are the main source of construction funds, and government bond investment is also mainly used for infrastructure construction. However, the use of government bond funds is strict, and the effectiveness of supervision over the use of funds is poor. At present, although the legal person system for construction projects has been implemented, the government is still the main source of construction funds. At the same time, the government's monopoly on infrastructure investment is not conducive to competition, resulting in low investment efficiency and weak supervision over the use of funds. After China's accession to the WTO, the entry of foreign equipment and management technology is conducive to the construction of infrastructure and the marketization of management. Collective and private entities account for more than half of domestic deposits, making it possible for private funds to enter the public infrastructure sector. However, the government's "market failure" and low service quality in the management of BOT projects already implemented will affect economic development, and the implementation of public-private partnership PPP mode is conducive to the solution of this problem.

2. The implementation of PPP mode is conducive to the government to reduce financial pressure and make up for the shortage of public funds. China has a large demand for infrastructure investment. As an important supplement to public project investment, social capital can relieve the pressure of public project investment and reduce the financial burden. The country has invested more energy and financial resources in national education, science and technology, and the construction of social security mechanism.

3. The implementation of PPP mode is conducive to improving the effective supply of public project services and the level of public services. If all public projects are monopolized by state-owned companies, they lack competitive pressure and motivation; have high operating costs, low investment, and insufficient supply; and cannot meet the growing social needs. The introduction of private capital to moderate competition, private competition to promote monopoly competition, can increase supply. Competition lowers prices and enables consumers to obtain services with high cost performance. For example, after the government in the wireless communication service, public transportation and water supply industries relaxed the market access restrictions, a large amount of private capital was invested to set up, and the supply quality was improved to a certain extent, thus greatly improving the city's communication, transportation, and tap water supply. Of course, the public sector needs to set up service quality standards, establish an effective incentive and supervision mechanism, and promote the private sector to enhance its technological innovation capability and improve its management efficiency and service level.

4. The implementation of PPP mode is conducive to reducing operating costs and realizing the best value of funds. Due to the profitability nature of the private sector, its commercial sensitivity is used to promote social investors and the public sector to fully control risks and costs, avoiding the public sector's cost-free operation. Using the technical and management advantages of social investors, we will establish a creative and cost-saving public service provision method to provide public services at a lower cost.
5. The implementation of PPP mode is conducive to controlling project risks. The risk will be borne by the department best suited to identify, resist, and manage the risk, and some risks will be transferred to private units that can be better controlled. For example, the private sector will bear cost overruns, on-time service, and market risks, while the public sector will bear the responsibility of purchasing public products, controlling service quality, and taking bad policy risks to protect public interests. Therefore, the risk of public projects can be effectively controlled by reasonably sharing the risks through public-private cooperation, giving full play to the advantages of both public and private parties, and properly integrating the market with the government.
6. The implementation of PPP mode helps to change the government functions. The cooperation between the government and non-governmental investors should be negotiated to reach a contract to regulate the responsibilities and rights of both parties. If the government breaches the contract, enterprises can sue or withdraw funds and facilities. If the enterprise breaches the contract, the government can also sue or take over the management of the franchise project ahead of schedule and start other competitors. The government has focused its main efforts on the formulation of rules, policy guidance, and supervision according to law. At the same time, the government will evolve from a simple executive commander to a cooperator, enhancing the awareness of cooperation with society and enterprises and promoting the continuous improvement of social integration capability.
7. The implementation of PPP mode is conducive to the development of capital market. China has great potential for social investment. More than 20 trillion social savings deposits can unleash huge investment power. Creating conditions to guide and transform social capital to directly invest in public infrastructure industries and taking advantage of the clear property rights of social capital and the strong self-binding nature of investment are conducive to reducing the pressure on commercial banks and preventing financial risks.

With the gradual integration of government and market, public-private partnership has become an important institutional choice in the process of marketization of public projects. The government uses the operating efficiency and competitive pressure of private organizations to improve the production efficiency and technological innovation speed of public projects. At the same time, the government uses cooperation contracts to regulate the behavior of private companies and realize the public welfare goals. Private companies can also use contracts to protect their own interests and avoid losses caused by the discretionary power of regulatory agencies. In order to adapt to the process of open competition in public infrastructure industry, various measures should be taken to encourage private capital to enter the field of public project investment. Through the implementation of PPP mode, social investors and government departments should play their due roles in public infrastructure industry.

1.2.6   Implementation Problems Faced by PPP Mode

The implementation of PPP mode faces several problems: (1) How can the effectiveness of local government's guarantee for private companies be solved through contracts and laws? (2) the financing policy of private institutions; (3) How to solve the problem that local governments cannot issue debts to support projects? (4) How to strengthen the management of private supply of public goods in order to improve the supply efficiency of public goods? According to the report of the World Bank, there are two types of management of private supply of public goods in countries all over the world: one is countries with strong institutional capacity, and the management of public goods supply is generally limited in price. The second is that countries with weak institutional capacity usually adopt a management commitment mechanism for private investors in public goods and use a third-party (such as the World Bank) guarantee to solve the commitment guarantee, so as to avoid noncommercial risks to private investors and lenders, including government confiscation and nationalization. China is in the process of economic restructuring. On the one hand, the government's administrative control ability is strong; on the other hand, the government's ability to rule by law is weak. Therefore, it is unrealistic or inefficient to attempt to rely solely on formal regulation and management, and the use of various management commitment mechanisms may have more advantages than disadvantages.(5) China's current PPP model does not clearly define the risk transfer mechanism, and the government does not guarantee the return on investment of PPP projects. However, in the actual operation of PPP projects, there are already three investment return modes, namely "self-financing investment return mode," "fixed ratio investment return mode," and "flexible ratio investment return mode." In addition to "self-financing investment return mode," the government has guaranteed the project company's investment return rate. "Fixed ratio" refers to the amount of investment return borne by the local government of the project according to the investment return rate determined through prior consultation. Fixed ratio support by the government for PPP projects is the result of Nash equilibrium. The essence of "government support" is to transfer the project cost from the consumers of the project to all taxpayers. Therefore, the government should carefully choose the supported projects. The purpose of government support is to attract a large amount of private investment for such projects and promote their success. (6) The current PPP model has not yet formed a perfect bidding system. For example, the bidding operation is not standardized; some bidders resort to deceit, collude in bidding, snatch the bid at a low price, or seek the bid by bribery and other improper means; the department has too much administrative examination and approval and intervention in bidding; etc. As the reform of investment and financing system, PPP mode is the first step in the reform of public project construction system. Under the current PPP mode, how to increase the maximum payment value VFM of national tax, that is to say, how to provide low-cost and high-quality public services for national tax, is a problem of system innovation. After private capital entered the public sector, private enterprises changed from public space to power space, thus making private capital a new monopoly capital. How to prevent dynamic private capital from becoming monopoly capital in organizational structure. The innovation of PFI system in British public sector tells us that as a kind of privatization strategy, a "synergy" is formed by reshaping the functions of government and non-governmental organizations under the situation of serious shortage of national financial resources. How to reduce the power space and expand the public space in the public domain is the primary problem of public sector system innovation. PFI's inspiration to China is: how can the top-down public goods supply system dominated by the national market be transformed into the bottom-up public goods supply system dominated by the people? It not only meets the fair value but also meets the efficiency value, reduces the expenditure scale of public finance, and provides low-cost and high-quality public services.

Other problems facing the implementation of PPP mode: including the choice of cooperative companies; The responsibilities and risks that the government should bear in cooperation; PPP model involves a large number of interest units and complicated organizational relationships. It requires high management level of participants. How to establish an effective management model and decision-making mechanism is worth further study. How to set the project return rate is a controversial issue in PPP projects. How to solve the policy risks and credit risks faced by the operators, as well as the construction of supporting measures and legal systems in accordance with the franchise management measures, etc.

Although the history of PPP in our country is not very long, many problems have been exposed in the previous practice of PPP mode.

1.2.6.1   The Government Functions Were Unclear

Although the cooperation agreement stipulates the rights, obligations, and responsibilities of both parties, due to the imperfection of the third-party supervision mechanism, the government is still in a strong and advantageous position in the process of investment and operation of specific projects. The government is both a referee and an athlete. If the government breaks its promise and defaults, it will bring direct damage to investors' financing costs and project income. One is that when the government treats the relevant private sector, it is easy to encourage private sector investment when it is needed, and it is unwilling to continue cooperation when it is not needed, or to change the investment policy in disguised form to urge the private sector to withdraw. Changchun municipal government sewage treatment project is a typical case of this problem. In order to properly handle the sewage treatment problem in the city, Changchun municipal government has decided to jointly invest with Huijin Company in the construction of a Sino-foreign cooperative sewage treatment plant and issued the "Changchun Huijin Sewage Treatment Exclusive Management Measures." The management measures specify the operation mode of this project and the rights and obligations of all participants and are one of the basic documents for the cooperation of this PPP project. However, by 2003, the Changchun municipal government had abolished this management measure on the grounds that it violated the State Council's regulations on "fixed returns." After that, Changchun Drainage Company stopped paying the sewage treatment fee, resulting in a total of 97 million yuan of overdue sewage treatment fee, causing serious losses to Huijin Company. In the end, the project ended with a government buyback.

The second is the lack of perfect preliminary planning and budget constraints. In order to introduce social capital, the government generally promised to provide better subsidies and demand guarantees at the beginning of the project. However, when problems occur in the actual operation of the project and it is even difficult to sustain it, the government often fails to fulfill its promise, and due to the lack of effective constraints, the final failure of the project becomes inevitable. The root cause of this kind of problem lies in that some local governments only regard PPP mode as a financing mode, but do not pay enough attention to its essence of improving public service efficiency. Therefore, in order to attract capital and reduce short-term financing pressure, the government often makes promises that are out of actual bearing capacity, which can only end in government dishonesty and project failure.

1.2.6.2 The Lack of Relevant Legal System

The lack of PPP-related legal system has increased the transaction cost and operation difficulty of social capital. PPP projects are generally closely related to public life, and their quality and price are issues of public concern, which are also easy to cause public opinion to rebound. However, up to now, China's PPP model still lacks a legal system at the national level. There are only lower-ranking departmental rules, local rules, and other normative documents in the entire legal system. Such documents are generally formulated by ministries and commissions of the State Council or local governments according to their functions and powers. Looking from the horizontal direction, its content lacks overall consideration and has great regional differences. From the vertical point of view, in reality, the formulation of the aforementioned documents is mostly driven by policies, and the revision and change of policies are relatively frequent, thus causing the aforementioned contents to change frequently. In addition, due to the lack of perfect and powerful supervision mechanism in the formulation process of the aforementioned documents, the contents of such documents are more added to the will of the formulation department and even deviate from high-ranking legal documents such as the constitution, laws, and administrative regulations. All these make the recipient department bear extra risks and costs in PPP projects, which is not conducive to the development of PPP projects.

1.2.6.3   Operating Mode Was Simple

At present, the application of PPP mode in China is still limited to the field of infra-structure, while in some countries or regions with relatively mature PPP develop-ment, such as Britain and Australia, PPP has long been widely used in the field of public services with smart public facilities, such as education, health, justice, public housing, etc. In addition, the implementation period of PPP mode is too short to give full play to the role of PPP mode. Currently, some PPP projects (such as BT projects) implemented in our country usually have an operation period of only 3–5 years. The short implementation period not only violates the original intention of the government to solve the financial financing problem but also increases part of the cost due to the involvement of the private sector, which eventually leads to the loss that outweighs the gain. For example, in some places, the government usually pays 40% after the construction of the city loop line through PPP and then pays another 30% every year for 2 consecutive years. As the financing cost of enterprises is generally higher than that of the government, the project will not only fail to leverage the financial capital but also further increase the debt burden of the government.

1.2.6.4   Lack of Reasonable Risk Sharing Mechanism

PPP projects generally have a long duration, and neither party can predict and bear risks independently. Therefore, PPP agreements must be active and must have adjustment and price adjustment mechanisms. If social capital takes on too many operational risks and cannot obtain reasonable returns, the project will inevitably be difficult to operate continuously.

For example, in 2003, Hubei Shikan started privatization of public transportation in the whole city. Private entrepreneur Zhang Chaorong accepted 68% of the shares of Shiyan Public Transportation Company with 38.16 million yuan, and his com-pany paid 8 million yuan to the government every year to obtain the franchise of Shiyan Public Transportation Line for 18 years; however, with the gradual increase of operating costs, the company's profit margin has been severely squeezed, but government departments have banned the increase of bus fares and refused to give subsidies, resulting in great difficulties and contradictions in the operation of the project. Eventually, the project had to be terminated, and the government had to take over the management again. Beijing Metro Line 10 has a relatively successful risk sharing mechanism. After the agreed fare for the project is determined, the annual assessed passenger flow shall be taken as the basis for profit distribution. If the pas-senger flow is 90–110% of the contract scope, the risk of passenger flow change shall be borne by the investors. If the passenger flow is lower than 90%, the govern-ment will subsidize the passenger flow to 90%, if the passenger flow is higher than 110%, the corresponding profit of the excess will be divided equally by both parties. If it exceeds 120%, the profits corresponding to the excess will be divided between investors and the government. The terms of the contract also stipulate a mechanism for adjusting profits to foreign exchange, raw materials, quality, price, and inflation rate.

It can be seen from this that problems such as unreasonable risk allocation and rigid benefit allocation in PPP projects have caused great losses to the private sector. However, if the government departments bear too many risks and responsibilities, the PPP project model is just a superficial one and fails to realize its real meaning. Generally speaking, China's PPP projects are often accompanied by various financial concessions, such as financial subsidies and tax concessions. Although these are nominally PPP models, the government still bears some responsibility for covering the bottom, which often increases the government's burden.

1.2.6.5 The Lack of a Reasonable Coordination of Management Agencies

Inadequate supervision and lack of project experience are also important reasons that have restricted the development of PPP before. There is no special PPP management organization in China before 2014, and there is no corresponding project development and reserve. The successful implementation of a PPP project requires a long period of preparation, usually 2–3 years. At present, the lack of value for money evaluation system and the unclear understanding of the applicable types of PPP projects in China have directly led to the blindness of choosing PPP mode in China. For example, the promotion of Qingdao Veolia Sewage Treatment Project is due to the government's unfamiliarity with PPP cooperation, the government's uncertain attitude leading to a long negotiation time for the contract, and due to the lack of an effective evaluation mechanism in the market; although a contract was signed on the sewage treatment price, the government later requested to renegotiate and reduce the price. Shanghai Dachang Waterworks Project is due to the high fixed rate of return on investment set in the contract design, which violates the laws of the market, seriously damages the public interest, and ultimately leads to the failure of the project. China's PPP practice since the 1990s is mainly a spontaneous exploration from bottom to top, mostly limited to the project level, and lacks experience summary, institutional arrangement, and theoretical guidance. Therefore, it is urgent for the central level to set up a PPP working mechanism, promote the establishment of a PPP management institution, and strengthen unified guidance on risk sharing, competition mechanism, reasonable government commitment, follow-up contract management, etc.

Hartwich et al. believe that the factors that affect PPP include common interests, costs and benefits, synergy efficiency, conflicts, and interest sharing. In the process of cooperation, identification of cooperation opportunities, formation of common interests, commitment, and agreement are all necessary conditions for the success of PPP mechanism construction. In addition, the importance of different factors is not the same. Based on CGIAR's PPP cooperation project, a study of 42 participants from different institutions shows that the participants' cooperation expectations and cooperation risks have a greater impact on PPP, while the cost-benefit and common benefits have a relatively small impact, which is consistent with Hartwich's research results on Latin America. The latter's research shows that although cost-benefit is advocated as the decision-making principle in cooperation, few public research institutions carry out cost-benefit evaluation on cooperative research projects before and during actual cooperation. On the one hand, the lack of responsibility caused by the failure of the principal-agent mechanism in the public sector is the cause of insufficient cost-benefit assessment; on the other hand, the project assessment is difficult, the assessment cost is high, and the accuracy of the results is poor.

The cooperative efficiency of PPP is one of the important bases for participants to make cooperative decisions. In 2007, the corresponding efficiency evaluation criteria were established: $E(CI)pr+IA\ pr \leq E(B)pr$ and $E(CI)pu+IA\ pu \leq E(B)pu$. Among them, $E()$ represents expectation, CI represents transaction costs of participants, IA represents cooperative investment, B represents cooperative benefits, and pr and pu represent private and public R&D participants, respectively. When the expected cooperation income is greater than the sum of the expected transaction cost and input cost, the participants will choose cooperation. The cooperative efficiency of PPP requires participants to effectively manage cooperative projects. Rowe et al. combed the common principles of 11 international PPP management agencies, including the ability of partners, complementarity of partners, responsibility and transparency, fairness and openness of project selection, honest communication, consistent and clear objectives, public welfare of results, trust and cooperation, extensive cooperation, strategic planning and long-term funding of partnerships, and identification and management of potential legal and ethical issues. These principles reflect the potential influencing factors and existing problems in PPP cooperation.

Management principles are conducive to the standardization and scientization of management, but what cannot be ignored is that managers' own preferences may have a significant impact on PPP. Hartwich et al. believe that the decision-making of public departments is mostly made by managers, and the decision-making results are more likely to reflect the personal preferences of managers rather than the needs of the public, leading to the deviation of public research subject objectives. Spielman et al. showed that the behavior of decision-makers or managers directly affects private expectations. Public institutions that leave negative impression on private individuals such as slowness, inefficiency, and conformism usually bring negative expectations to private individuals, thus leading to failure of cooperation. Therefore, maintaining a high degree of coordination between public and private sector managers in PPP is one of the key points for successful cooperation. The sponsors of PPP are also decisive in the cooperative participation of public research institutions. As the funds for agricultural research and development PPP come from the competitive project funds of government departments, fund managers have indirectly decided which institutions can participate in PPP when formulating the standards and conditions for the use of funds, which may result in unequal opportunities for different institutions to participate in PPP and is not conducive to the exertion of efficiency advantages.

1.2.6.6   Lack of Financing Channels

At present, China's PPP projects mainly rely on bank financing, and bank loan financing methods often need guarantee or mortgage, and the financing channels are relatively narrow. In the traditional bank credit business, innovation may be needed in the PPP field. In November 2014, the State Council issued the "Guiding Opinions of the State Council on Innovating Investment and Financing Mechanisms in Key Areas to Encourage Social Investment" (Document No. 60 for short), which explicitly mentioned supporting innovative loan businesses such as emission rights, charging rights, collective forest rights, franchise rights, expected benefits from purchase service agreements, pledge loans for collective main land contractual management rights, etc. In addition to traditional credit financing channels, the experience of other countries shows that PPP projects can raise long-term low-interest funds from institutional investors such as pension funds and insurance companies. The system reform of the Chinese government in recent years has laid a certain foundation for promoting the abovementioned changes in financing methods, but the new financing mechanism still needs to be tested and applied.

1.2.6.7   Exit Mechanism Was Not Perfect

The withdrawal mechanism of PPP projects in our country is still far from perfect. If it is not for the state-owned enterprises to undertake PPP projects, it is almost impossible for private enterprises to profit from PPP projects and then withdraw smoothly, because the withdrawal in the middle needs to be approved by government departments and private enterprises need to find suitable substitutes. Both at home and abroad stipulate that investors' withdrawal must be approved by the government. Failure to do so will easily lead to short-term speculation by enterprises. Moreover, a proper exit mechanism can stimulate investors' enthusiasm.

1.2.7   Suggestions on Perfecting the Financing Mode of Public-Private Partnership in China

1.2.7.1   The Laws and Regulations Perfect

The successful implementation of PPP project cannot be separated from the protection of a mature legal system. Perfecting relevant laws and regulations system is the necessary premise to ensure the development of PPP mode in China. On the one hand, it is necessary to speed up the formulation of top-level design and enact laws at the national level to regulate the rights and obligations of all parties involved in PPP mode and to clarify their responsibilities and risks, so that local governments have laws to follow when formulating policies. At the same time, it also regulates the uneven policies in different places. On the other hand, each local government should form a set of specific and complete laws and regulations, which should be scientifically stipulated in terms of government authorization, market access, tendering and bidding, project application and approval, financing channels and methods, performance guarantee, land acquisition, use and income, ownership and transfer of specific rights and interests, information disclosure, dispute resolution, exit mechanism, etc., so as to improve the transparency of policies, let the private sector and the public understand relevant laws and regulations and ensure the smooth operation of PPP mode.

In PPP financing mode, private partners, as the partners of infrastructure construction, are in a disadvantageous position compared with government capital. Therefore, the restriction of contract norms is their rights and interests to seek self-interest protection. This kind of guarantee is multilevel and includes the distribution of rights and interests in many key steps such as contract formulation, contract signing, and contract implementation. If there is no legal protection, PPP financing mode will face great difficulties, and the financing efficiency will be extremely difficult to improve, which shows the legal perfection of financing guarantee. At present, many governments have already adopted laws, regulations, and rules related to franchising. However, from a macro perspective, the laws or decrees promulgated are at a lower level, lack of cross-industry universal protection, and are not easy to enter other regions or regions to explore high costs. This kind of insufficient financing protection is very unfavorable to the promotion of PPP mode. In addition, these provisions in all aspects are too brief in content, many problems have not explored the evidence of the law, and many problems are lack of practical operability, and only stay at the level of principle formulation.

PPP financing projects are subject to cross adjustment according to various laws, of which contract law is the main basic basis for adjustment. The forms of both parties' participation in the contract gradually play an important role in the adjustment of interest distribution. For PPP financing mode, its legal background is complex, and the financing parties include the government, investors, and international organizations, which will cause the legal applicability problems caused by the identities of different actors. China's urbanization speed is faster and faster, and the scale of infrastructure construction is also accelerating. The fertile soil and needs of domestic infrastructure construction have attracted more and more capital participation of large-scale enterprises, so as to obtain the revaluation of enterprise capital through consortium investment behavior. The current PPP financing mode requires important technical guidance. When conducting investment negotiations abroad, it is necessary to pay attention to the legal provisions that may lead to conflicts and abide by international practices. In a word, our country is short of legislation with practical significance at present. The administrative legislature needs to conduct multiparty investigation and mainly analyze the market prospect, so as to stimulate the new development of the economic situation.

1.2.7.2 Strengthen the Government's Sense of Responsibility, Improve the Project Risk Sharing Mechanism

Learning from the development experience of PPP model in western developed countries, the establishment of scientific risk sharing mechanism has become one of the necessary links in the application of PPP model. Especially for the private sector, risk factors are important factors for their cooperation with the government. A reasonable risk sharing mechanism can attract enough social funds to participate in the construction of PPP projects. The government needs to clearly divide the risks of all parties in the project design and formulate corresponding measures to deal with the risks. Resolutely avoid taking full responsibility for the project design, taking risks into one's own arms, and avoiding responsibility by wrangling with each other after problems arise. Only by sharing risks, benefits, honesty and trustworthiness, mutual trust, and win-win, the smooth development of PPP projects can be guaranteed.

The traditional government has various functions, which bear many functions such as investment, project construction, later operation, and maintenance. The cost in construction projects is very high, and the overall operation efficiency is low due to multifunctional coordination. The above situation has seriously affected the development of public infrastructure and the improvement of urbanization level. Therefore, the Chinese government needs to carry out reforms in terms of functional transformation. It should not delegate all powers to the construction of projects. Instead, it should play a main role in macro-control of project planning, policy formulation, project supervision, and other aspects to promote the reform.

1. For the overall layout of the project construction, government departments are the main makers. At present, the current situation of China's infrastructure still has many difficulties, such as overall shortage and lack of construction funds. The introduction of PPP financing is just to ease the pressure of fiscal expenditure and to share the risks of infrastructure investment costs with private capital and international capital. Government departments need to improve their service level, adhere to the planning and design of infrastructure based on laws and regulations, and cannot make various adjustments to policies as leaders like to ensure the basic interests of the partners in a long cooperation cycle.

2. Infrastructure can be subdivided from the perspective of realizing public service functions, roughly in the following three categories. The first type is private consumption of construction project funds, such as utilities, transportation, and other services. This type of project can collect fees from the public to recover costs during later operation. Policy guidance and subsidy support are needed for basic projects that cannot recover costs. The second category is public consumption, or social services, such as municipal transportation, green space construction, etc., which cannot be charged after completion and need financial power support and subsidies. The main investor is the government department. The third category is located between private consumption and public consumption,

including sewage disposal and park construction. After the construction of such projects, some fees can be levied on specific users, but financial subsidies and discount interest production are still needed. Generally speaking, although there are some differences in the degree and proportion of subsidies, government subsidies for public infrastructure projects are essential.

3. Pay attention to the interests of the public and enterprises, and strive to be the intermediary coordinator. For enterprise capital invested in infrastructure projects, the fundamental starting point is profit. The characteristics of infrastructure projects determine that consumers need to enjoy the long-term quality service of infrastructure projects. The convergence point of this contradiction between profit and function is charge. When the two sides coordinate, the government can optimize the basic pricing mechanism, improve the functional diversity of project services, ensure the convenience of project layout, and ensure the operation of construction projects.

4. To strengthen the government's supervision and supervision of infrastructure projects in terms of investment, construction, and operation, to promote the marketization of investment, to optimize the decision-making procedures coordinated by the partners, and to scientifically and reasonably ensure the income balance of the decision-makers, the government departments also need to carry out appropriate guidance and financial discount to balance, optimize the stress rate of the investment budget, and guide social capital into infrastructure construction. Further, it is also possible to build a good decision-making management, distribute financial revenue well, and improve the attraction of financial leading projects to social capital. Currently, PPP is the best multiparty cooperation mode with government investment as the main body. This model can improve the efficiency of government functions to maximize and liberate the use of administrative forces, better play the government's supervision function, and protect the actual interests of investors, and the public interest cannot be better satisfied. If government supervision is not effective, the damage to one party's interests will continue to affect the ultimate interests of corporate capital and financial funds.

1.2.7.3 Create Conditions and Create Various Financing Modes

The exploration and improvement of financing mode need to be improved from three angles:

1. The establishment of private capital guarantees organization. Investment funds for infrastructure construction are huge, and all come from bank credit. These huge capital demands make capital turnover difficult. Obviously, it is very necessary to set up a special organization to provide guarantee services for private capital. In order to create a good PPP financing environment, it is also necessary to establish credit files in a timely manner and standardize the administrative

agencies that guarantee and to solve the problem of principal financing for existing guarantee institutions and guarantee quotas.

2. We should actively promote the system reform of government management. Infrastructure projects need the active involvement of the public sector. The public sector needs to use public-private cooperation to ease financial expenditure, and foreign capital enters infrastructure projects through PPP financing mode when participating in projects and carrying out institutional investment management. As a matter of fact, investors can only take the initiative to invest in construction after they know that their own interests can be effectively protected. They can also actively participate in PPP project financing activities and more actively look for new PPP infrastructure investment opportunities.

3. It is necessary to expand financing methods and improve the diversity of project financing channels. For this reason, the banking department should optimize the approval process for PPP financing, improve the efficiency of capital utilization, improve the traditional conservative concept, and scientifically and reasonably provide loan support. However, most of this support comes from state-owned banks. For commercial banks, there is still a defect of low enthusiasm for such financing. Therefore, a capital risk operation market should be established, and investors should be provided with bonds, stocks, and other financial instruments to raise the necessary construction funds.

1.2.7.4   Public-Private Partnership Financing Project Satisfied with Price Adjustment Model

The main participants of PPP financing mode consist of three basic types: the first is government agencies, the second is private enterprises, and the third is the public. From the perspective of pursuing the interests of all parties involved, the objectives of the three parties are not the same. Government agencies focus on how to improve the attraction of PPP financing, increase the enthusiasm of enterprises to participate in infrastructure investment, and compensate for the historical difficulties of financial shortage through reasonable guidance of private capital. Government agencies are under capital control and will reduce government subsidies and subsidies from contractors as much as possible to alleviate the preferential treatment of construction capital. For private enterprises and private capital, its essence is to pursue the reporting rate of investment. Its project management objective mainly focuses on whether it can obtain sustainable policy support and benefit protection from government departments and at the same time obtain more steady returns in the operation after the completion of the project. The public is the experience and enjoyment of the project results, and a large part of the financial expenditure comes from public taxes. Therefore, the public is both an investor and a result tester. They mainly pay attention to the diversity, convenience, and appropriate investment burden of the project services. From the above statement, the objectives of the three parties involved are more independent, and in fact there are also more overlaps and intersections. The unity of the three objectives is to find a suitable way to meet the

Fig. 1.3 Endogenous
cycle diagram

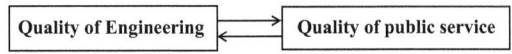

objectives of both private enterprises and the public and to be more in line with the objectives of government departments. According to the principle of system dynamics, the internal relations among the three are studied, and the target cycle of PPP financing is confirmed as follows:

As shown in Fig. 1.3, the increase in the proportion of public efficiency will lead to an increase in the budget, and at the same time, it will also increase the passenger flow and increase the operating income of private charging enterprises. This dynamic operation makes the feedback always in a balanced state. When the above balance state is stable, all three parties involved have positive expectations on the feelings of the target. On this basis, this paper also designs the adjustment of the feedback mechanism for the target system, focusing on the price setting and adjustment of public services. This positive dynamic price adjustment mechanism is conducive to the public and private owners to reach their goals.

1.2.7.5 To Strengthen Personnel Training

PPP franchise project is a kind of project with both opportunities and challenges. Its successful implementation and construction require enterprises to have many positive foundations and factors. First of all, participating enterprises need to have a thorough understanding of PPP mode and know all detailed procedures well. Improving the level of investors at the decision-making level will be beneficial for enterprises to carry out reasonable and scientific feasibility studies. At the same time, enterprises also need to understand and control the risk control and fund recovery at various stages of project planning, construction, implementation, operation, and management.

PPP financing projects are managed and operated worldwide through bidding. For PPP financing, its projects need to involve many interdisciplinary and complex fields such as financing, financing methods, taxation, project implementation, investment estimation, project decision-making, relevant laws, etc. The demand for comprehensive talents such as economy, management, and technology is increasing, which requires enterprises to consider in the long run, increase the training of compound talents, train high-end talents with practical pioneering spirit, and further fully guarantee the implementation process and normal operation and management of the project.

1.2.7.6 Strengthen the Social Credibility and Ensure the Government's Credibility

Both the private sector and government departments must have a good social reputation to ensure the efficient operation of PPP projects. On the one hand, the spirit of the contract should be strengthened. After winning the bid, the private sector should sign a contract with the government with clear powers and responsibilities. In case of unilateral default by the private sector, intentional delay, or suspension of work, the government has the right to initiate the recovery procedure and publicize it to the public, punish the dishonest acts of the private sector, and establish its integrity awareness. On the other hand, government departments should strengthen the construction of information publicity, improve the performance evaluation system, and improve the social supervision system to promote its social credibility. It is beneficial to win the trust of the private sector, thus promoting the better development of PPP mode.

1.2.7.7   Establishing a Reasonable Price Mechanism to Protect Public Interests

The "publicity" of public goods and services requires government departments to pay attention to the realization of public interests. The PPP mode is also a means to improve the supply efficiency of public goods or services, and its ultimate purpose is to protect public interests. Therefore, government departments must put public interests first. And a reasonable price mechanism is an important link to protect public interests from damage. When setting prices, government departments must carry out a scientific and reasonable hearing system, presided over by government departments and involving various stakeholders such as the private sector, the public, experts, and scholars, to ensure that prices are scientifically set under reasonable cost calculation. For some PPP projects with good social benefits but poor economic benefits, the government should give reasonable financial subsidies. On the premise of stable prices, ensure the normal operation of the private sector for the project.

1.2.7.8   Dealing with the Relationship Between Fairness and Efficiency, Emphasizing Fairness

Dealing with the relationship between fairness and efficiency has become an important topic in China's economic construction. In particular, the construction of public infrastructure, because of its own public goods, has a strong "public," so it is more necessary to put fairness in the first place. This requires government departments to carry out reasonable supervision and guidance in many aspects such as scheme design, price setting, and operation supervision. Especially for regions with relatively backward economic level and people with relatively weak purchasing power, they are often neglected by the private sector. Government departments need to meet their needs for basic facilities through corresponding subsidy mechanisms and preferential policies and to protect their basic rights and promote the fair development of society.

1.2.8 Theoretical Research on Public-Private Partnership Model

1.2.8.1 The Theoretical Basis of Public-Private Partnerships

The public goods theory divides products into three categories from the perspective of consumption characteristics: pure public goods, quasi-public goods, and private goods. Pure public goods are personal consumption of such products or services, which will not harm anyone else and benefit from their consumption. The typical characteristics are non-exclusive and noncompetitive. Therefore, pure public goods are usually provided by the government, not for efficiency reasons but because the private sector cannot provide pure public goods (or the cost is too high), or cannot overcome the "free-rider" problem. The government refers people to overcome market failure. Private products are entirely provided by private individuals. Quasi-public goods are more common than pure public goods. Both the government and the private have the responsibility to supply them. The competition and exclusiveness of quasi-public goods are strong or weak. Therefore, different modes of supply must be adopted according to their characteristics.

Agency theory is rooted in institutional economic theory. After the public service reform in western countries entered the "new public management reform" stage, under the promotion of the normative rationality of public sector behavior and the financial department and relevant government departments, based on the concept of minimizing transaction costs and reducing costs in the company theory, and applying the traditional economic behavior assumption in the public choice theory to government decision-making, the market mechanism and modern competition theory were introduced to the public sector, thus expanding the agency theory, destroying and developing the PPP model, and providing new ideas for reshaping the government structure.

In PPP projects, the competition mechanism in the bidding market forces bidders to lower their costs. Cash prices are transferred from the private sector to the public sector. At the same time, the private sector is responsible for the whole process of project design, construction, operation, and maintenance and bears the risks of design, construction, and operation. In addition, the private sector also faces government credit risks, technological risks, and interest rate risks. It can be seen that with PPP mode, the public sector can obtain a better cash price and avoid the related risks than it can supply public products by itself.

The PPP model was first proposed by the British Conservative Party and can be accepted by the Labor Party and developed vigorously. It is then widely popular in the West and has an objective basis for its existence and development.

- *Government service consciousness.* The relationship between the government and the market is clear. The government serves the market and taxpayers, thus generating the internal reform impetus for the government to promote the efficiency of public expenditure. On the other hand, the government is also the representative of the public interest. It emphasizes efficiency while giving consideration to fairness. Only in this way can it serve the public better.

- *The development of private economy.* In PPP mode, most of the partners of the public sector are the domestic private sector. It not only has financial strength but also should have talents, technology, and good management level. It is the material basis for PPP mode to be implemented.
- *The market economy is orderly.* The "partner economy" is also a "contract economy." The cooperation between the two parties is based on good faith. The private sector cannot abandon vicious price competition in terms of service quality and level. The government cannot gain a strong negotiating position as a manager. Moreover, it cannot artificially depress the discount rate and squeeze the reasonable profits of the private sector. In particular, the matters not specified in the contract and the influence of some unpredictable factors are all settled on the basis of good faith.
- *Public sector management level.* PPP model is not an empirical paradigm but a scientific system management project. Not only do they have rational expectations of the economy and society, they also need scientific analysis methods and tools. For example, in the London Underground PPP project, regulating the relationship between contract subjects involves more than 300 mathematical formulas and numerous legal provisions, requiring the government to have a fairly high level of management. In addition, the public has the expectation to obtain high-quality public services, and there is a strong demand to reduce the tax burden, while the government is facing pressure to limit debts. The implementation of PPP mode has political, economic, and other environmental foundations.

1.2.8.2 Theoretical Research on Public-Private Partnerships

1.2.8.2.1 The Perspective of the Relationship Between the Contract Theory

PPP is essentially a contract between the public and private sectors to jointly provide public services. The contractual relationship in legal sense is a subset of the social relationship voluntarily undertaken. As shown in Fig. 1.4, the transactions of

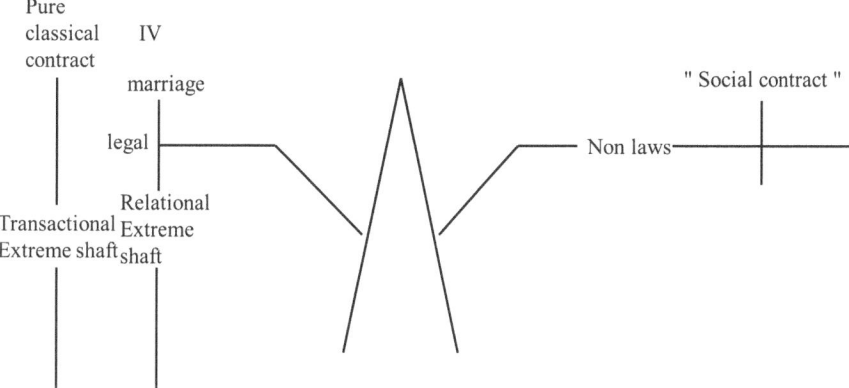

Fig. 1.4 Contract type continuum

the "extreme axis of transaction" are completely protected by law. As the degree of "relatedness" of contracts increases, they are less and less protected by law and rely more on traditional or internal execution tools.The analysis of relational contracts can be carried out in two ways. One is to use formal optimization models to show the consequences of rational behavior along the standard microeconomics theory, which is called "incomplete theoretical contracts." The other is carried out in a descriptive way and adopts the assumption of limited rationality, which is called "relational" contract theory. Relational contracts are only protected by law to a limited extent, and nonlegal measures are taken to a large extent to realize future uncertainties and possible opportunistic behaviors. Therefore, both parties should establish some form of governance mechanism to realize cooperation. Of course, they are not completely dependent on self-performance. Some form of third-party governance such as regulation can also improve the implementation effect of contracts.

Essing believes that PPP is a contractual relationship between the public sector and the private sector. Both parties share skills and assets, share risks, and share the potential benefits of providing public services. According to the contract theory and the organization theory, PPP has the characteristics of relational contracts and is more dependent on formal cooperation based on trust, as shown in Fig. 1.5.

Bovaird believes that PPP is a new type of partnership and alliance, often without any legal support or restriction. Although it also includes contractual support, it often realizes partnerships that exceed contractual commitments. After a detailed comparison between the transaction contract relationship and the partnership

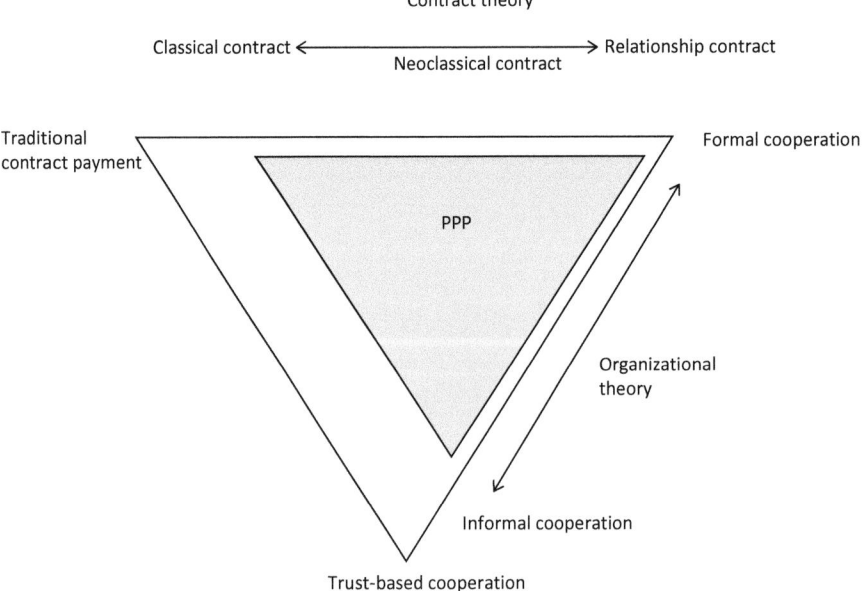

Fig. 1.5 PPP contract nature

Table 1.4 Governance principles of governance contracts and partnerships

The principle of governance	Transaction contractual relationship	Partnerships
Citizen participation	Consult with citizens and other stakeholders	Participation of citizens and other stakeholders in decision-making
Transparent	Limited to areas that stakeholders need to know in order to supervise the agreement, even this may be affected by trade secrets	Publicize all participants, including users and citizens, as an important factor in building trust
Responsibility	The agreement must follow uniform procedures, especially those related to budget and cost control	Be responsible for any dispute that may arise and be prepared to be responsible for the overall performance of the entire partnership
Equality and social inclusion	Only when the agreement was clearly defined	As a core value of the partnership, and seek innovative ways to improve performance
Ethics and honesty	Act in line with the law, in line with professional standards	As a core value of the partnership, and seek innovative ways to improve performance
Fair (fair procedure and appropriate process)	Must be treated in accordance with the organizational procedures, consistent with everyone	As a core value of the partnership, and seek innovative ways to improve performance
The willingness and ability to cooperate	Subject to attention but not necessary	Is an extremely important factor for all partners
The ability to compete	Extremely important factors for service providers	Extremely important factors for the overall partnership
Leadership	Good organization and management is necessary (timely, accurate, and effective to meet the requirements of the contract)	In every organization that forms an organization or service, all levels of partnership are important
Sustainability	Must meet all the sustainability criteria set by the contract	There is still a need to improve methods to improve the sustainability of policies and activities

relationship (Table 1.4), it is considered that PPP research does not include the inter-organization relationship that is managed, supervised, and implemented based on the detailed provisions of traditional contracts, although these relationships are sometimes called "partnership." Grimsey believes that PPP is a long-term contract and cannot include any changes in the environment. It is not a transaction contract and should be regarded as a relationship contract. Contract management requires specific technologies.

In short, the understanding of PPP in relational contract theory highlights two points: first, PPP contract is not a classical transaction contract but a relational contract, which is a typical incomplete long-term contract; and secondly, the implementation or governance of the contract is important, and specific technologies and mechanisms are needed to ensure the effect of the contract.

1.2.8.2.2 The Perspective of Transaction Cost Economics

Starting from the assumption of bounded rationality, transaction cost economics holds that all contracts are incomplete contracts, takes transactions as the basic analysis unit of the study, holds that different transactions produce different transaction costs, and confirms three main factors that affect transaction costs: transaction frequency, uncertainty, and asset specificity. Production costs include technical costs and management costs (administrative coordination costs), comprehensively compare production costs and transaction costs, and adopt corresponding governance structures (market, mixed, bureaucratic, and bureaucratic organizations) according to the principle of minimizing production costs and transaction costs (governance costs). The main analysis framework is shown in Fig. 1.6. It is believed that the transaction cost is mainly determined by the asset specificity. According to different asset specificity degrees, the production cost and management cost of two different discrete structures are compared (Fig. 1.7). When the most asset specificity level (K) is small, the market procurement has advantages. When the optimal asset specificity level (K) is high, the internal organization has advantages; when the optimal asset specificity level is medium, mixed governance will occur, and PPP is one of the mixed governance forms.

The important concept of transaction cost economics is asset specificity, and the focus of contract management is to prevent opportunistic behavior caused by asset specificity. Essig et al. believes that the way the government provides public services should be based on the consideration of asset specificity and strategic importance, as shown in Fig. 1.8. When the asset specificity and strategic importance of the product are low, the transaction cost is low, and the government adopts the purchase method; when the asset specificity and strategic importance are both high, the transaction cost is high, and the method directly provided by the government is adopted; when asset specificity and strategic importance are between the above two, the government can adopt PPP. Parker et al. studied the economics of PPP from the

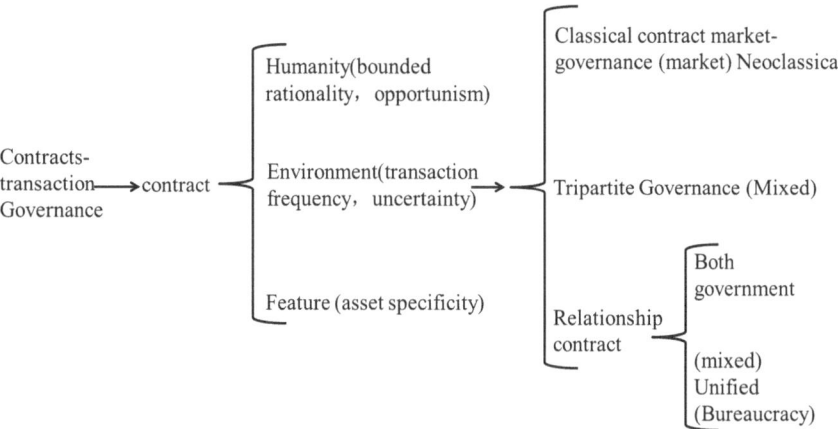

Fig. 1.6 Transaction cost economics analysis framework

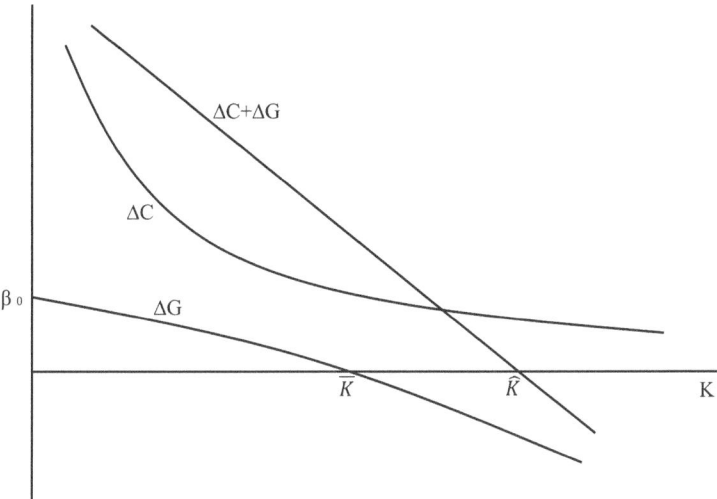

Fig. 1.7 Governance costs and asset exclusivity

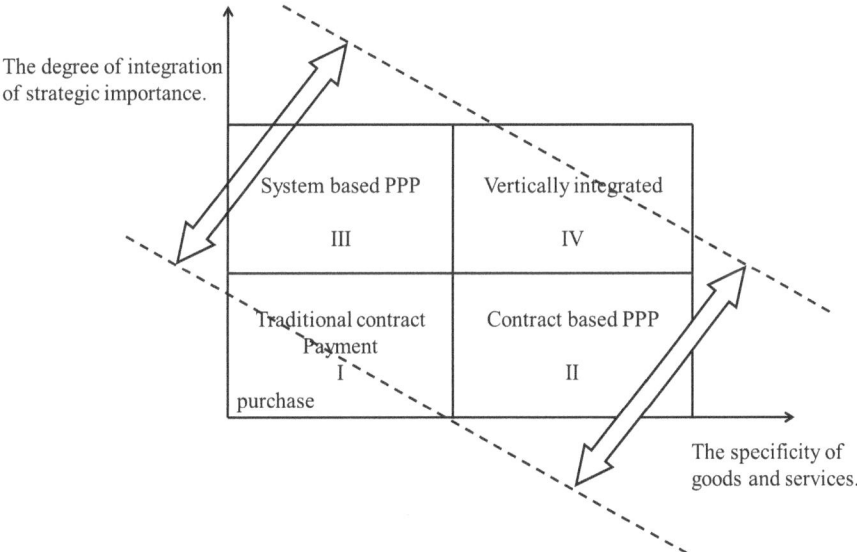

Fig. 1.8 PPP applicable interval based on asset specificity and strategic importance

perspective of transaction costs and especially analyzed the role of transaction costs and the importance of trust in relationship contracts. Firstly, the paper reviews the transaction cost literature from the perspective of the resource base of purchase decision-making and then considers the role of trust and reputation from the perspective of minimizing transaction cost. Given that information asymmetry, asset specificity, and opportunism may harm contracts, the paper emphasizes the role of transaction cost and holds that the transaction cost theory, supplemented by scope economy and scale economy, provides a powerful analytical framework for the study of government procurement policies and PPP.

In a word, transaction cost economics takes bounded rationality and opportunism as assumptions, starting from the incompleteness of contracts, and holds that the key is to adopt different governance structures to minimize transaction costs. According to the principle of minimizing production costs and transaction costs, it is believed that PPP is an ideal form for the government to provide public services under certain circumstances. Trust, cooperation, and commitment can reduce transaction costs and are important to the governance of PPP.

1.2.8.2.3   The Perspective of Property Rights Economics

Property right economics is another branch of incomplete contract theory, which holds that property right represents the residual control right and claim right of income. Due to bounded rationality, all future contingencies cannot be included in the contract, so property rights are important. Due to bounded rationality and opportunism, incomplete contracts and exclusive investment are lower than the optimal level. Therefore, it is required to assign ownership to the important or indispensable party in the investment.

Hart proposed a preliminary PPP model and believed that the key feature of PPP is that the construction and operation of public service facilities are tied together and illustrated by taking prison management as an example. If the prison PPP project is divided into two stages of construction and operation, the choice between service outsourcing and PPP is very simple. Without binding, builders have neither internalized social benefits nor internalized operating costs, making as little productive investment as possible. In the case of binding, builders do not internalize social benefits but internalize operating costs, resulting in increased productive and nonproductive investments. The conclusion is that whether the two tasks of construction and operation are bound, that is, whether PPP is adopted, depends on whether the results of the two tasks are easily defined and measured. When the quality of service facilities is easy to define and the quality of service provided is not easy to define, it is more suitable to use the traditional "unbundling" method. On the contrary, if the quality of service facilities themselves is not easy to define, and the quality of service provided is easy to define, it is suitable to apply the "binding type," that is, PPP method. Besley et al. analyzed how property rights work when the private sector participates in the provision of public services, i.e., which property rights arrangement results in the largest joint surplus. The results show that if the contract is incomplete, the property right should be given to the party whose property right income is higher in value, regardless of the relative importance of investment or other aspects of production technology, and the attribution of the most property right should be analyzed from the two dimensions of investment productivity and output value factor of public and private sectors, as shown in Fig. 1.9.

Combed the research on PPP property rights economics and divided the typical PPP into four stages: design, financing, construction, and operation. The key characteristics of PPP are service outsourcing, private investment, and task binding. PPP

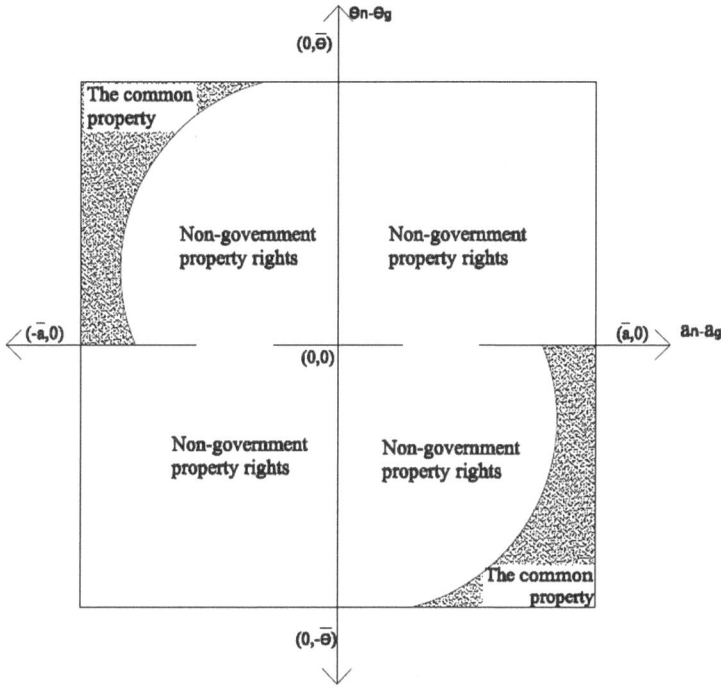

Fig. 1.9 Optimal allocation of equity in the provision of public goods

gives property rights to the private sector to realize high-energy incentives, but whether PPP is implemented or not is related to the specificity of investment, environmental certainty, transaction cost, observability, and measurability of product quality, whether output services are public goods, and whether various tasks are complementary. A classification framework of public service provision modes is proposed, as shown in Fig. 1.10. It is believed that the mode of service provision is mainly based on asset specificity and environmental uncertainty. When asset specificity does not exist, market provision is adopted. When there is asset specificity, if the environment is complex and uncertain at the same time, and the transaction cost is high, the government will provide it. If there is no environment complexity and uncertainty, PPP-type long-term contract will be adopted. Martimor et al. also believes that the key feature of PPP is the binding of "construction" and "operation" tasks. First of all, under the condition that a complete contract can be written between the public sector and the private sector, from the perspective of mutual externalities between the two tasks of "construction" and "operation," and taking maximizing social welfare as the criterion, the efforts of the private sector under the two situations of task binding and no binding are compared. When the efforts of the private sector on both "construction" and "operation" tasks are greater than those without binding in the case of task binding and PPP, the task binding method will be adopted, otherwise it will not be adopted; in addition, under the framework of

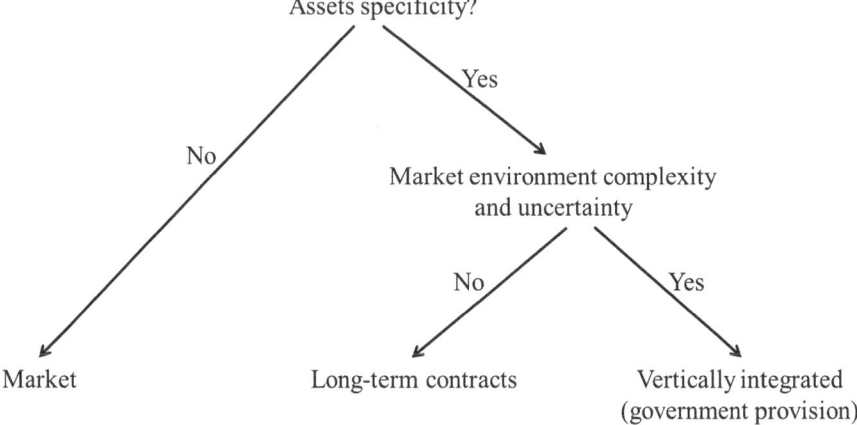

Fig. 1.10 Classification framework for public service provisioning

incomplete contract, the optimal usage conditions of task binding and unbundling are analyzed.

Bole and others believe that property rights are important, but property rights alone are not enough. Strong supervision and governance are needed. They are classified according to the degree of control and distribution of property rights (Table 1.5). It is pointed out that the recent privatization (PPP) is characterized by strong control of private sector property rights while they are owned. Estache et al. studied the introduction of private sector reforms in the energy, water (water supply and sewage treatment), and transportation sectors in developing countries and countries in transition and found that the impact of property rights on efficiency improvement is different between the water and transportation sectors. In the field of transportation, the private sector in developing countries usually performs better than the public sector. However, for the water industry, most cross-border studies have not found significant statistical differences between the efficiency of the public sector and the private sector, and the role of property rights does not seem as large as commonly believed.

In a word, the essence of property rights economics is to study how to achieve the maximum output through the optimal distribution of property rights in various parts of PPP, focusing on the prior distribution of property rights to achieve high-energy incentives, ignoring the implementation and governance of contracts.

Table 1.5 Classification of public services based on ownership and control

The degree of control	Ownership	
	Public	Private
Minimal control	Most of the public services before 1980 (all public services of the Soviet Union)	Most of the "typical" privatization of the 1980s
Strong control	Recently tested as a substitute for privatization	Recent privatization

1.2.8.2.4 The Perspective of Game Theory

Game theory mainly studies the process of decision-making interaction between two or more people. PPP is the result of decision-making interaction between the public sector and the private sector. Game theory can be used to analyze and study PPP. Scharle regards PPP as a social game, thinks that analyzing PPP phenomena and experiences from the perspective of game is conducive to a better understanding of PPP, and selects four perspectives to discuss, namely, linguistics, formal theory of game, experimental psychology, and system theory. Evidence shows that analyzing the rules, roles, results, strategies, and payments of PPP governance process from the perspective of game theory can better understand its failures and difficulties. The social game emphasizes that PPP is a real cooperative undertaking. The public sector has key legal and moral assets, while the private sector brings external capital, technology, and incentive structures, which are essential to the success of PPP. In addition, the utility, risk characteristics, and influencing factors of game participants are analyzed, and it is believed that the utility function of many PPP participants is not a linear function of money, especially the utility function of government staff has its own characteristics and is very risk-averse. Even if it has high returns, it is unwilling to take the lead in action. Moreover, the results are often controversial due to political aspects, so the risk sensitivity of government staff also exists in mature democratic countries and is not equivalent to the political instability in less-developed countries. In addition, risks, costs, and benefits must be reasonably distributed, and the multiple interests of multiple participants must be skillfully coordinated. The application of game theory makes it possible to clarify unclear and uncertain situations, viewpoints, and actions in the PPP game process.

Ho developed a game model to analyze the renegotiation between the government and the private sector in PPP, believing that the government often renegotiates with developers in order to save PPP projects that are experiencing difficulties. When a developer of a PPP project requests government funding, the government should compare the advantages and disadvantages between the bankruptcy of the project caused by the refusal of funding and the grant of funding, thus causing many problems in the negotiation and management of the project, providing a framework and method for studying the dynamic behavior of all parties in PPP. The model is shown in Fig. 1.11. Besley et al. actually used the perspective of game theory when analyzing the role of property rights in PPP. He believed that the important role of property rights is to enable the exclusive investor to have greater bargaining power. Even if the cooperation fails, the exclusive investor can still obtain a higher reservation effect.

In a word, adopting the perspective and method of game theory in the research of PPP can deeply understand the process and results of PPP. However, due to the strict assumption of game theory, the utility of public sector in PPP is difficult to quantify accurately, and the application of game theory in PPP research is still preliminary.

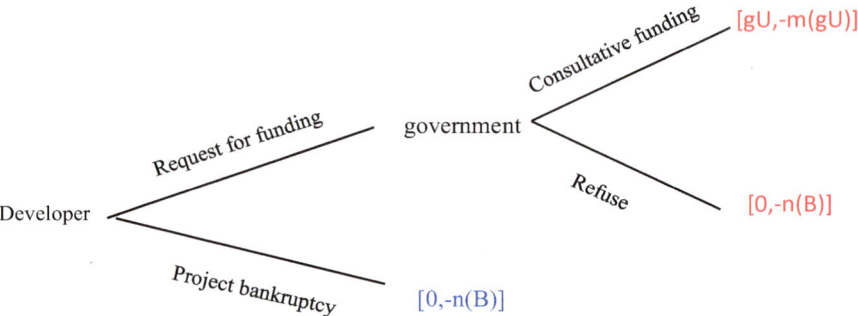

Fig. 1.11 Game model of PPP project renegotiation. (Source: Ho 2006, p. 680)

1.2.8.2.5   Review of Existing Research on the Theory of PPP

Private sector investment in PPP is a highly specialized asset, which can be explained by transaction cost economics. The provision of public services can take two forms, directly provided by the government or PPP. The former has low transaction costs but high production costs, including high technical and administrative costs. The latter has low production cost but high transaction cost. By comparing the sum of production costs and transaction costs under different governance structures, and deciding whether to adopt PPP, it also shows what kind of institutional environment and governance mechanism PPP needs. However, transaction cost economics cannot explain a key problem in PPP, that is, how to distribute the benefits generated by public-private cooperation between the government and the private sector. Classical transaction cost economics analyzes whether a product is purchased in the market or produced by the enterprise itself, and if it is produced in the enterprise, it will yield benefits. Then the income is entirely owned by the enterprise, and there is no problem of distribution with other organizations. PPP involves the cooperation between the government and the private sector. The success of the cooperation depends not only on whether the cooperation generates benefits but also on the distribution of the cooperation benefits between the two sides. In other words, the cooperation may not be realized due to the high cost (sum of production cost and transaction cost) or may fail due to the distribution of cooperation income. However, transaction cost economics obviously cannot explain this very well, so the application of transaction cost economics alone cannot provide a complete analysis framework for PPP.

The school of property rights theory represented by Hart holds that due to bounded rationality and opportunism, contracts are incomplete, resulting in exclusive investment below the optimal level. Therefore, it is required to assign ownership to the important or indispensable party in the investment. According to the property rights theory, the main problem of incomplete contracts is the incentive distortion of prior exclusive investment caused by the rip-off of time. Therefore, the emphasis is on prior incentive, denying the transaction cost afterwards; holding that "the negotiation cost is zero"; holding that there is no difference between the market and the enterprise in terms of incentive intensity, administrative control, and adaptability; and holding that property rights are the most important and the key lies in who should merge and who owns all material assets. However, this is not consistent with the situation of PPP. PPP is not only a transfer of property rights within a certain period of time but also a typical long-term incomplete contract. The rate of renegotiation is very high, which requires a large transaction cost afterwards and requires a high governance mechanism. Therefore, the property right theory can provide some insights for PPP, but it cannot fully explain the PPP problem.

The concept and principle of strategic alliance have been introduced into some research related to PPP governance, which is believed to be conducive to better sharing risks, promoting innovation, and improving management skills. However, the strategic alliance itself has not yet formed a unified theoretical explanation, and there are important differences with PPP. Although the strategic alliance theory can provide some ideas for PPP governance, it cannot establish a complete analytical framework for PPP.

1.2.9 Analysis of Factors Affecting the Performance of Nine Public-Private Partnerships

1.2.9.1 Bargaining Power

In PPP contracts, the partners obtain Pareto optimality, but the contract results have many different Pareto optimality situations and benefit distribution methods. The actual results reflect the bargaining power of the contracting parties, including negotiation power, threat power, and patience. For example, Andersen found that when studying Danish government's provision of dental care services to the public, the balance of power between public and private sectors plays an important role in service results such as coverage rate and stability. Of course, it may also mean that one party gains the greatest benefits at the expense of the other party's interests. Therefore, Wettenhall believes that the ethics and legitimacy of public services must be maintained in PPP to effectively protect public interests.

There are many factors that affect the bargaining results, including the tactics adopted, the negotiation procedures, the information structure, and the discount rate of the participants. Yan thinks that bargaining power can be divided into two categories: bargaining power based on environment and bargaining power based on resources. The former is mainly reflected in the importance of negotiation itself and its results to negotiators and other options, while the latter is mainly reflected in the amount of key resources that both parties have or control. Inkpen further pointed out that knowledge, professional skills, and information can become important bargaining power. Specifically in PPP, the private sector's proprietary knowledge, expertise, and information are important bargaining power. Therefore, the bargaining power of the public sector is affected by its own ability and information. According to studies by Romzek and others, the stronger the competition, the more professional knowledge, and the more transparent the information, the more favorable it is for the government to bargain. Therefore, the bargaining power of the government in PPP is determined by the following two aspects: (1) whether relevant personnel in the public sector have sufficient professional knowledge and skills, such as accumulation of experience and training of relevant personnel, etc., and (2) whether the public sector can obtain information, such as through the participation of external consulting agencies, consultants, stakeholders, etc.

1.2.9.2 The Retention of Utility

The reservation effect of the public sector in PPP is reflected in the original public service level before the private sector enters. If the private sector is allowed to enter into the operation of the existing infrastructure (stock assets) through PPP, the public sector's reservation utility is mainly reflected in the efficiency of the direct operation of the former government before the private sector enters. If new infrastructure (incremental assets) is built and operated through PPP, it is also closely related to the public service level before PPP appeared. If the infrastructure construction has a large funding gap, the government's financial situation is poor, and it is unable to provide necessary public services, the retention effect is low. The situation in most developing countries and countries in transition is that, on the one hand, the traditional way of direct government provision is inefficient and difficult to provide public services that satisfy the public; on the other hand, the government is short of financial resources, unable to provide sufficient public services, unable to update the existing infrastructure, and unable to sustain development. For example, in China, on the one hand, the government is facing a shortage of funds for infrastructure construction, renovation, and maintenance; on the other hand, the operating efficiency of the existing infrastructure under the traditional planned economy is unsatisfactory. In general, the public sector's reservation effectiveness in PPP in transition countries is low, which is also an important reason for the transition countries to vigorously develop PPP in reality. Therefore, we take the low reservation utility of the public sector in the transition countries as the established premise of the study and mainly study the reservation utility of the private sector.

The private sector mainly invests capital, experience, and professional skills in PPP. The main sources of private funds in PPP projects are as follows. First, private sectors own funds, especially those enterprises that are in the forefront of the industry in the world have abundant funds, and a considerable proportion of project funds may be their own funds. Secondly, equity financing is the exchange of part of the equity of the project management right for the injection of funds or other assets. Since the investors of PPP projects are mostly nonlocal investors, investors often seek partners who are familiar with local conditions to improve project operation efficiency. However, due to the high rate of return required by equity, equity financing often accounts for only a small proportion of PPP project financing, and most of the funds come from creditor's rights financing, including bank loans, consortium and institution loans, and corporate bonds. Bank loans are divided into mortgage loans and secured loans. The former generally refers to obtaining bank loans by taking the cash flow of the project's future operating income as a mortgage, and loans can also be obtained through pledge of charging rights when the project obtains charging rights. The latter refers to the project company's shareholders or other guarantors providing guarantees for the project company to obtain loans. The consortium loan process is often simpler, but there are usually additional conditions, which can be chosen when the project is risky, the income is low, the bank loan cost is high, or the local bank is unable to lend. Institutional loans apply for noncommercial loans from policy institutions such as the World Bank. These loans often have low interest rates and many preferential terms, but they have some specific requirements for financing projects and are difficult to obtain. Corporate bonds generally issue corporate bonds with the future income of the project as collateral, and the future repayment of principal and interest also comes from the operating income of the project.

Most PPP projects in developed countries use project financing. Project financing refers to a financing mode that takes the future net cash flow of the project and the project's own assets as the guarantee to repay the loan and is characterized by project orientation, limited recourse, and risk sharing. Project financing is a modern financing mode, which is obviously different from the traditional corporate financing mode in financing concept (Table 1.6). At present, project financing has been widely used in all parts of the world. Its sources of funds are more extensive, the duration is longer, the government intervention is less, and projects are involved in all industries. Project financing was introduced into China in the mid-1980s, but due to various reasons, the development of project financing in China is relatively slow.

The financing methods of PPP projects in transition countries are quite different from those in developed countries. Take China as an example. First of all, government equity participation is relatively common. Enterprises with government background often exist in the equity structure, resulting in the government taking part in some market risks while participating in projects through its affiliated enterprises. However, in large foreign PPP projects, the government hardly takes any market risks. Secondly, bank loans are mostly limited to domestic banks. And they are mainly state-owned banks, which are greatly influenced by the national financial policies. The source of loans is single, and loans are often made to only one bank, resulting in the project's capital supply being completely affected by the operating conditions and strategies of one bank, thus increasing the project's financing risks. In addition, the proportion of corporate bond financing is small, while foreign PPP projects can usually solve the project capital demand by issuing domestic bonds.

The utility of private sector reservation is mainly manifested in the cost of funds or the difficulty of obtaining funds. If the cost of funds is low, both parties will benefit greatly, and PPP will have a high probability of success. For example, according to the statistics of the World Bank, the main reason for the peak of private sector investment in infrastructure in Europe and central Asia around 2000 was the sufficient capital in the international capital market at that time; in addition, the Beijing municipal government had planned to introduce a large number of private sectors

Table 1.6 Key factors for PPP success

The key to success	Secondary factors
Strong investment environment	A stable political system, a strong economic system, sufficient local financial markets, predictable exchange rate risks, a predictable and reasonable legal framework, government support, community understanding and support, the project is in the public interest, predictable risk options, the project itself is suitable for privatization, the economic situation is very good
Economic feasibility	Project products have long-term demand, limited competition from other projects, the profitability of the project can attract investors, long-term cash flow attractive to borrowers, long-term cooperation of suppliers for long-term operation of the project
With strong technical advantages, trusted franchise consortium	Leading role of key enterprises, effective project organization structure, strong and capable project team, maintain good relations with relevant government departments, having partner skills, rich experience in public-private partnership project management, multi-domain participants, good technical solutions, innovative technical solutions, efficient technical solutions, the environmental impact is small, pay attention to public safety and health
Good financing plan	Good financial analysis, investment, and payment plan; stable sources, structure, and currency; high equity/debt ratio; low financing cost; fixed and low financing rate; long-term debt financing with minimized risks; easy to handle interest rate fluctuations and exchange rate risks; appropriate charge level and adjustment formula
Appropriate allocation of risks through credible contractual arrangements	Appropriate and reliable risk sharing, franchise contract, stakeholder contract, design and construction contracts, loan contract, insurance contract, supply contract, operating contract

into the infrastructure construction for the Olympic games, but in the end it was mainly state-owned enterprises. The important reason was that the capital market was imperfect, the capital cost was high, and it was difficult for private enterprises to obtain sufficient capital in a short time.

1.2.9.3 Production Cost

The main purpose of establishing a partnership between the public sector and the private sector is to improve the efficiency of public service delivery by taking advantage of the low production cost of the private sector, while the production cost of the private sector is mainly determined by its technology and management level. Therefore, the technology and management level of the private sector have an important impact on PPP performance. For example, Field et al. found that it is very important for government departments to correctly choose strategic partners when studying PPP in British medical industry. Hofmeister et al. believes that an important aspect of the successful implementation of PPP is to carefully select partners and absorb experienced private sector. Zhang also believes that an important factor for the success of PPP is the public sector's selection of suitable private sector partners. The private sector must have strong capital, advanced technology, and high management capability. Zhang further confirmed that the private sector in PPP must have obvious technical advantages, a strong project team, and rich experience in project management. In short, the private sector's technology and management level determine its production costs, which in turn affect PPP performance.

1.2.9.4 Private Transaction Costs

We follow the classic transaction cost economics to analyze the transaction cost of the private sector. Transaction cost economics was originally used to explain the substitution relationship between enterprises' self-production and purchase from the market and then expanded to explain the different governance modes between enterprises and the market. In recent years, the analytical framework of transaction cost economics has been applied to analyze different modes of public service provision, such as PPP or direct government provision. PPP mode has low production cost, but the transaction cost between the private sector and the public sector is high. When the transaction cost offsets the benefits brought by the reduction in production cost, PPP becomes an inefficient governance mode. Therefore, the size of transaction cost has an important impact on PPP performance. If transaction cost is low, PPP performance is high, and if transaction cost is high, PPP performance is low. For example, Meinard empirically studied the transaction characteristics and transaction cost to determine whether the government directly provides public services or adopts PPP method and the performance level after adopting PPP method. Chong et al. confirmed the view that "high transaction costs lead to PPP inefficiency" through empirical research.

According to Williamson, the size of transaction costs mainly depends on asset specificity and uncertainty. Asset specificity refers to "the extent to which an asset can be reallocated to other alternative uses or used by alternative users without sacrificing its production value." This is related to the concept of precipitation cost. Williamson believes that asset specificity is the most important, and the importance of asset specificity to transaction cost economics cannot be overemphasized. It is generally believed that the transaction cost increases with the increase of asset specificity, but specific analysis should be made according to specific situations. For example, Hwang believes that the effect of asset specificity on transaction cost is influenced by environmental factors such as the trust level between organizations and the duration of cooperation between both parties. In addition, Lui et al. believes that unilateral exclusive investment will result in being "blackmailed" by the other party, increasing opportunism and transaction costs, while the mutual exclusive investment of both parties indicates the willingness and commitment of both parties to cooperate and creates obstacles to withdrawal. Both parties are more inclined to consider cooperation from the long-term interests and solve problems through negotiation, thus reducing transaction costs. Empirical research generally supports the discussion on asset specificity in transaction cost economics. According to David et al.'s statistics on empirical research of transaction cost economics, asset specificity is found to be the most common independent variable for testing. Sixty percent of testing supports the analysis of transaction cost economics, and only 4% of testing is significantly contrary to the prediction of transaction cost economics. Therefore, we believe that asset specificity has an important impact on transaction costs, especially in PPP, private sector unilateral private investment is often very high, which will bring high transaction costs.

Transaction cost economics also believes that uncertainty is also very important. Only when there is uncertainty can asset specificity show its significance. At the same time, uncertainty has an important impact only under the condition of strong asset specificity. If special investment is not involved, people can easily form new trade relations, and it is not worth maintaining the continuity of the contract. However, in the case of strong asset specificity, if the uncertainty is too high, it will lead to high transaction costs, making hierarchical governance more efficient, and mixed governance (including PPP) becomes inefficient. The common influence of asset specificity and uncertainty on the governance structure is shown in Fig. 1.12.

As for the mixed system (mixed organizational form), Williamson initially believed that it was an unstable intermediate transitional organizational form and later realized that it was a common organizational form, including franchise, inter-company partnership, strategic alliance, etc. Menard further explained mixed organizations, believing that the combination of asset specificity and uncertainty with important consequences leads to opportunistic behavior and disharmony, leading to the emergence of mixed organizations. If there is only one factor, short-term contracts are preferred. The mixed organizations are further classified (Fig. 13). Organizations close to the market end mainly rely on trust. Organizations close to integration mainly rely on government supervision; the relationship network in the middle has realized closer coordination than trust and has formal rules and traditions. However, taking leadership as the coordination mode, it mainly relies on the authority to supervise the partners relatively closely. For the governance of mixed organizations, there are mainly contracts, reputation, trust, adaptation mechanism, negotiation, and formal authority. In addition, the complementary investments of the participating members of the mixed organization will generate a quasi-"rent," and its distribution is crucial.

Rankan et al. specifically analyzed the conditions for public-private partnerships, as shown in Fig. 14. According to transaction cost economics and externality theory, PPP is considered necessary only under the following circumstances: first, when the realization of economic opportunities requires competitiveness related to specific industries and has important positive externalities for society, it means that specific private sector actions involve important public interests; secondly, when there is high uncertainty for the private sector, economic opportunities will be lost without public sector participation. In addition, the private sector faces high contract signing,

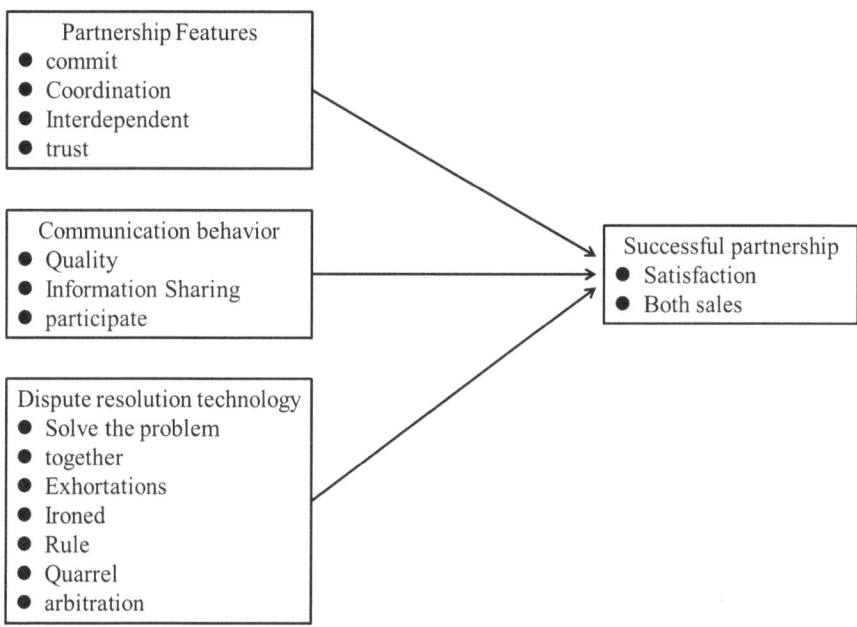

Fig. 1.12 Analysis framework for successful partnership

coordination, and implementation costs, and such governance costs are necessary. In short, specific resources, political externalities, uncertainties, and governance costs jointly determine the formation of a partnership. When public relations benefits are high, public sector resources costs are high, while when private sector resources costs are low and have resource advantages, both public and private sectors can generate benefits through joint actions. When the private sector's governance cost is low and the degree of uncertainty it faces is low, joint actions can take the form of contracts. Otherwise, the private sector needs to form a constructive partnership with the public sector. Since most PPP in reality is carried out in the form of contracts, it is very important to reduce the governance costs and uncertainties faced by the private sector.

Different scholars classified the uncertainty according to different standards. Williamson distinguished the environmental uncertainty causing external interference to the transaction from the behavioral uncertainty caused by speculation by both parties to the transaction. Grover reviewed the application of transaction cost theory in the field of operation and supply chain management and divided uncertainty into environmental uncertainty and behavioral uncertainty. The former mainly reflects environmental unpredictability, including natural, technological, and institutional environment, while the latter mainly reflects opportunistic behaviors of public and private sectors. Menard et al. divides uncertainty into internal uncertainty and external uncertainty. Internal uncertainty is mainly related to the input, output, and transformation process. The input-related uncertainties may be due to unobservable transaction resources or services, difficulties in coordination between inputs or lack of commitment from external suppliers, etc. The uncertainty of output does not match the consumer's preference due to the difficulty in setting standards or lack of flexibility in adapting to changes in demand. The transformation process itself may produce uncertainties, such as caused by the complexity of technology and human skills.

Many existing literatures study the uncertainty and risks in PPP and their sharing principles. For example, Jinchang et al. believe that risks in PPP have two general attributes of all risks: future uncertainty and potential loss. Therefore, the risk in PPP can be defined as the possibility that the uncertainty of changes in things will bring some potential losses to the risk taker during the whole process of signing and implementing PPP contracts. Regarding the principle of sharing, a typical example is Liu Xinequality (2006), who believes that the principle of risk sharing can reduce the probability of occurrence of control, loss after occurrence of risk, and the management cost of risk, and the party with the most ability to control and deal with risk will bear the risk. Sadka classifies risks in PPP into endogenous risks and exogenous risks and believes that the party that bears the risks should be determined according to the nature of the risks (Table 3.1).

There are many sources of uncertainty. Under certain circumstances, certain uncertainties may play a key role while others do not. Therefore, the key is to find those uncertainties that play an important role. Williamson's environmental uncertainty mainly refers to disturbances caused by "natural random behaviors," "unpredictable changes in consumer preferences," and "lack of communication." However, it is assumed that the system is given and known, which is consistent with the reality of relatively stable institutional environment in western developed countries. For example, Meinard studied the impact of uncertainty on contract selection and performance in the French water supply industry, mainly considering unknown factors that may affect water consumption or water quality, including climate, regional economic development, and future population changes, and believed that the institutional environment was relatively stable and consistent. As an important source of external uncertainty, institutional environment is often mentioned, but there is little research and analysis on it. Grover has systematically analyzed the application and research of transaction cost theory in the field of operation and supply chain management and found that there are few literatures on the impact of institutional uncertainty and behavioral uncertainty on governance structure selection and performance. One of the few cases involving institutional environment is the influence of Oxley on the structure of interfirm alliances. It is found that the weaker a country's protection of intellectual property rights is, the more inclined the interfirm alliances in that country are to adopt a hierarchical structure. Therefore, it is necessary to study the joint influence of institutional environment and transaction characteristics.

For countries in transition, there is no essential difference between the key challenge of developing PPP and other countries, that is, to reduce risks and uncertainties, but the sources of uncertainties and risks are different from those of western developed countries. On the one hand, the environmental uncertainty emphasized by Williamson is still very important, because most transition countries are experiencing rapid marketization, industrialization, and urbanization and are facing great environmental uncertainty. On the other hand, uncertainties in systems, laws, and rules are very important in countries in transition, because not only is it itself an important source of uncertainty but also because PPP is an incomplete long-term contract, especially during the execution of the contract, various unpredictable uncertainties may occur. The handling of these uncertainties requires certain rules. If there are no corresponding rules, it will bring about great transaction costs. Therefore, we believe that for countries in transition, the uncertainty in terms of system, law, and rules is a higher level of uncertainty and a second-order uncertainty than that referred to by Williamson in traditional transaction cost economics. An important aspect of PPP development in transition countries is to have the necessary and appropriate legislative basis. For example, North has always insisted that rules are important for understanding participation in game theory behavior. Williamson further said that changes in parameters will lead to changes in governance structure.

In short, the lower the environmental uncertainty and behavioral uncertainty, the lower the transaction cost of the private sector in PPP. The analysis framework of private sector transaction costs is shown in Fig. 15.

On the whole, there is little research on the transaction costs of the public sector. Nunn believes that public managers should evaluate the costs brought by public-private cooperation to local governments, mainly including four types of costs, namely, approval costs and the early costs required for government officials to approve projects; dynamic expenses, expenses for coping with changes during the construction of the project; incentive fees to attract private sector fees; and operating costs, costs of protecting and operating new infrastructure. By estimating the joint airport project in Texas with an investment of 63 million US dollars, it is found that government officials estimated 25% less before the project was approved. Therefore, local government managers and legislators should estimate these four costs accurately as early as possible and pay attention to the nature of their dynamic changes before deciding on a public-private transaction.

Ferris et al. when studying the impact of transaction costs on the way the government provides public services, believed that the government's transaction costs mainly come from the process of contract writing and supervision. When the terms defining the service requirements are consistent, the contract writing cost is relatively low, while the community with homogeneous population composition may have stable service demand preference, which can reduce the writing cost and is verified by taking the ethnic proportion of the community population as an index. Supervision cost depends on service characteristics and types of private sector. Supervision cost is high when service output is not easy to measure and high when cooperating with profit-making organizations.

Another application of transaction cost analysis in the public sector is Dixit. This paper mainly analyzes the transaction costs in the process of policy making and holds that in the politics of transaction costs, one party to a political contract is a citizen (individual or interest group organization), while the other party is a politician (individual or party) or an administrative organization (regulatory organization, etc.) and the political contract is a transaction promise of policies (or decrees) and votes (or donations). Compared with economic contracts, political contracts are more complicated and more difficult to implement. For example, at least one party in a political contract has multiple parties, the contract is more ambiguous, and many political commitments are not restricted by any external enforcement mechanism. In the politics of transaction cost, both political system and organization can belong to the governance structure, which is characterized by a variety of agency relationships, including a variety of principal-agent and so on, and analyze the agency cost of the political process.

The classic is Horne who creatively applied transaction cost economics to study the choice of organizational forms of the public sector. His basic view is that in the process of achieving social goals, effective public management should ensure that transaction costs are minimized. The focus of the study is on the legislature. It is believed that legislators and their voters are involved in the transaction. Legislators should be supported by the election, while voters should get private benefits from the legislation or reduce the private expenses brought about by the legislation. It distinguishes four major transaction costs and the costs of legislative decision-making and private participation, which refer to the costs of defining the decision-making of transactions and the continuous private participation in the legislative process. The cost caused by the entrustment problem refers to the cost caused by the political threat to the reliability of the transaction caused by the inability to entrust to future legislators. Agency cost refers to the cost of ensuring the effective implementation of the executor and the loss of not doing so; uncertainty risk refers to the cost caused by the uncertainty of legislation. The author applies this analytical framework to explain why legislators choose different forms of enforcement organizations in different environments. For example, legislators choose between courts, independent committees, and administrative departments in the enforcement of regulations. In particular, the author analyzes how legislators choose between state-owned enterprises, private sectors, and administrative agencies when providing public services. However, the author mainly inspects the basic experience of the contemporary western developed countries in public management and its reform, with the legislature as the main analysis object and the legislative process as the main analysis content, which is quite different from the countries in transition.

According to the above research, we believe that the transaction costs brought by PPP to the public sector mainly include two types: decision-making costs and agency costs. Decision-making costs are likely to occur in the process of contract negotiation, signing, execution, and termination, mainly depending on the following aspects: the conflict degree of objectives (objective integration ability), the government's professional ability, and the degree of information acquisition. Agency costs occur in the whole process of PPP contract integration, change, or termination. First of all, the whole PPP process needs administrative officials to complete the task, so the degree of corruption of officials affects the agency cost. Secondly, in order to ensure that the private sector completes the contract, the private sector actually becomes the agency for the public sector to provide public services, and various mechanisms need to be adopted to ensure the implementation of the contract. In addition, if stakeholders have the motivation to participate in the implementation of the contract, they can fully monitor the implementation and reduce the burden of government supervision departments, thereby reducing agency costs and decision-making costs.

Decision cost exists in the whole process of contract negotiation, execution, and conclusion. PPP contracts are highly complex and require public managers to have a variety of professional skills. Contract experts divide the typical PPP contract process into three stages: the feasibility evaluation stage, in which the public manager decides whether a specific service is suitable to be provided by contract; the contract implementation process, including bidding, bid evaluation, negotiation, and contract signing; and, finally, supervision and evaluation of public managers in the performance of the contract to see whether the responsibilities stipulated in the contract are fulfilled. Whether there is contract experience which affects the decision-making cost, experience accumulation and the existence of economies of scale in managing multiple contracts enable the government to continuously improve its contract capacity, thus making more effective decisions and reducing decision-making costs.

PPP contractual relationship is an area between political rights and economic markets, and there is no set of rules that are fully applicable or frequently applicable. A good government needs to build its capacity and carry out reforms at both vertical and horizontal levels. The former is a model that relies on authority to drive and guide the execution of contracts from top to bottom, while the latter is a model that is driven by negotiation when the public sector enters the market. The ability and professionalism demonstrated by the government at each stage of the contract signing process is a way to ensure that more enterprises participate in bidding. The signing of contracts requires human and information resources and cannot be treated in the way that government departments handle daily business. The government needs to form a project team including project director, technical experts, audit experts, and cost accounting and operate well. Successful PPP requires contract planning, market analysis, and forecast. The more forecasts the public needs and the more effective market analysis, the better the government can make a good deal for the public. In short, whether a contract can be saved depends partly on the efforts and investment of government agencies in establishing a contract management system. If such capacity is not created, it will cost a great deal in the long run.

Under the circumstance that conflicts of interest make it difficult to reach a collective decision, the decision-making cost may be very high. The more intense the conflict between private interests that are closely related to decision-making, the more difficult it is to reach an agreement on decision-making. For example, it is often found in groups with different interests of internal members that it is easier to reach an agreement on the necessity of action than to reach an agreement on the action goal. If the decision-makers have strong heterogeneity, large differences of interests, or conflicts of interests, it is difficult to reach an agreement on the service objectives, and the decision-making cost is high. However, homogeneous groups often have stable and consistent service demand preferences, which are easy to form decisions and have low decision-making costs.

There is a complicated agency problem in PPP, which is twofold: on the one hand, the public sector acts as an agent for the public, and, on the other hand, the private sector acts as an agent for the public sector to provide public services. If these agency problems cannot be effectively solved, Pareto improvement for the private sector, the public sector, and the public cannot be realized.

Inconsistency of objectives refers to the different purposes and motives of the public and private sectors in PPP. The private sector pursues profits, while the public sector aims to safeguard public interests, which leads to high agency costs. Asymmetric information between the public and private sectors also leads to high agency costs. Therefore, transparency is generally regarded as a good factor in policy formulation. The greater transparency, the more accurate and symmetrical information and the less transaction costs, and even some transaction costs disappear.

Product measurability refers to the degree of difficulty in measuring service results or supervising service output activities in the public sector. Easy-to-measure services generally have ready-made, easy-to-confirm methods that can accurately measure the quality and quantity of services. If the result is difficult to measure, but the activity of supervising the output service is relatively simple, the service can still be easily measured. When the results and the activities that produce the results are not easy to determine, the service is difficult to measure, which will aggravate the opportunism tendency of the private sector, leading to the difficulty of supervision and high transaction costs in the public sector.

The government's regulatory capacity includes the setting of regulatory agencies, the professional ability of regulatory personnel, and regulatory means. The stronger the regulatory capacity, the lower the agency cost. Government departments can selectively use regulatory means, including paying attention to citizens' complaints, conducting citizen satisfaction surveys, analyzing performance data, and conducting on-site audits. It is necessary to effectively restrict regulatory discretion and solve the problem of supervision of regulators. The participation of the public and other stakeholders can realize continuous supervision, reduce costs, and alleviate the problem of information asymmetry. Authorization of stakeholders and their participation in supervision can improve the credibility of the commitments of all parties, realize the transfer of responsibilities, and prevent rent-seeking behavior of managers. This method of saving political transaction costs is especially important in developing countries, because the political systems of most developing countries lack the restriction and balance of opportunism in the public sector.

1.2.10 Research on Key Factors of Successful Public-Private Partnerships

1.2.10.1 The Perspective of Public Governance

Governance is a new development of the ruling mode, in which the boundaries between public and private departments and their internal boundaries tend to be blurred, with the ultimate aim of creating conditions to ensure social order and collective action. Governance involves multiple stakeholders, including but not limited to a set of social public institutions and actors of the government. The specific characteristics of governance are the fuzziness of boundaries and responsibilities among stakeholders, the interdependence of powers, and the autonomous autonomy of the network of actors. PPP is a way for the government to realize efficient public management and provide public services. Many research literatures on public administration have analyzed the factors that promote the success of PPP from the perspective of public governance. Bovaird believes that public governance is the interaction of stakeholders in order to influence policy outcomes, and partnership, as an integrated part of public governance paradigm, has more solid theoretical support. PPP should not only help to achieve public sector goals but also operate according to the principle of "good governance." Hofmeister et al. believes that the successful implementation of PPP requires both public and private parties to overcome traditional dogmas and paradigms and to govern PPP according to new rules and models. According to Switzerland's experience in implementing PPP, the following factors are considered to be extremely important to PPP's success: PPP must be carefully planned, long-term and short-term objectives must be balanced, and short-term behaviors must be avoided, especially the adoption of PPP solely due to financial difficulties; all relevant aspects must be considered in PPP project design, especially the rights and obligations of each partner must be clearly defined, and the incentive mechanism must be supplemented by effective enforcement and sanction mechanisms. Careful selection of partners, absorption of experienced third parties, support for public sector identification, and selection of private partners; fully reflect the country's political culture and other specific environment; project objectives must be clearly defined, and both parties must reach a balance on performance expectations to ensure fair agreement and fair risk sharing. Risks and opportunities related to projects need to be evaluated in a transparent manner and their development carefully monitored. In addition, successful PPP requires not only new rules but also theoretical models, spanning the distance between different disciplines and management tools, and evaluating the PPP performance that has been achieved.

Bloomfield believes that although PPP can theoretically benefit the public, including service quality improvement and cost reduction, the challenges faced by PPP long-term contracts jeopardize its successful implementation, including obstacles to market competition, reasonable risk sharing, performance guarantee, and appropriate transparency. If the government wants to implement PPP successfully, it must invest in specialized skills and effective contract management and perfect the governance structure. Field et al. studied the obstacles to the development of PPP and the factors that promote the success of PPP in the medical industry in the United States; pointed out the limited ability of the public sector to enter PPP; mainly reflected in the selection of suitable contracts and strategic partners, the correct prediction of the future environment, the negotiation, management, and supervision of PPP contracts, etc.; and stressed that the public sector should improve the corresponding ability. Brewer et al. believes that systematic trust is the key to establish and support the healthy development of PPP, which is also the principle encouraged by "good governance." He also studies two cases in Hong Kong and believes that the government's actions have seriously damaged the trust necessary for public-private partnership. One of them destroyed the trust due to the acrimonious public debate on the independence of service providers and the other because of inconsistent government policies and strong mutual suspicion between the public and the government, which led to a serious decline in the trust of community stakeholders and the public in the government, thus affecting the development of PPP projects. Therefore, in order to establish a good partnership, the government must carefully consider the principle of "good governance": responsibility, responsiveness, transparency, fairness, and participation.

Awortwi discusses from the opposite side that merely transferring the provision of public services to the private sector without ensuring that the basic elements for PPP success are in place will lead to worse situations than those reported in the existing literature. The study found that many supporters of PPP did not point out the abilities that public managers should have in understanding, negotiating, implementing, and regulating. They believed that PPP itself could not guarantee efficiency and reduce costs; contract implementation, contract supervision, competition, and local government capabilities were the core of PPP. The results show that PPP itself is not the cause of failure. The problem is that the local government has no ability to become a smart actor and adopts PPP hastily without new methods to solve public problems such as regulation, supervision, and promotion, which often leads to disappointing results.

In a word, the discipline of public administration is mainly from the perspective of public governance. A good public-private partnership should be characterized by cooperation, trust, transparency, and fairness. The government should improve its professional skills, enhance negotiation skills, correctly select partners, and strengthen supervision and management of PPP contracts.

1.2.10.2 Project Management Perspective

Jamali believes that although PPP is based on the comparative advantages of the public and private sectors, it still needs to consider the characteristics of both sides on a case-by-case basis, such as the need for commitment, interdependence, individual excellence, and communication. Therefore, trust, openness, and fairness are the basic conditions for the success of PPP. In addition, PPP must start with careful preparation and basic work. Public sector personnel must have professional skills and the ability to negotiate with the private sector on an equal footing. Developing countries must establish a good legal and regulatory framework and have strong institutions to guide and implement policies. Aziz et al. thinks that there are two main ways for the government to implement PPP: based on financing method, the purpose is to use private capital to meet the needs of infrastructure construction; and based on the service method, the aim is to achieve the best time and cost efficiency in service provision. The implementation of PPP may be hindered by laws, politics, and culture. For example, the US federal government has enacted a series of laws to promote the development of PPP in infrastructure. According to the detailed analysis of PPP projects in Britain and Canada, the principles that must be met when implementing PPP at the project level are obtained, including the existence of PPP legal framework and implementing agencies, understanding of private sector investment objectives and capital value objectives, keeping the PPP process transparent, and having standard procedures and performance definition standards.

Zhang believes that the key factors for PPP success include a strong investment environment, economic feasibility, technological advantages of the private sector, a good financing plan, and appropriate risk sharing through reliable contractual arrangements (Table 3.2). The public sector must select suitable private partners, and the private sector must have abundant funds and advanced technology, be able to guarantee safety and health and environmental requirements, and have high management capabilities. Zhang confirmed six aspects that hinder the success of PPP through a questionnaire survey: social, political, and legal risks; unfavorable economic and commercial conditions; inefficient public procurement mode; lack of mature financing and engineering technology; issues related to the public sector; and issues related to the private sector. It is believed that a perfect infrastructure PPP contract must include nine key aspects: the appropriate role of the government; realize the maximum value of funds; effective management consulting services; express an appropriate PPP plan; establishing relationship contracts; improving the purchasing mode; perfect the payment structure; defining the rights of contract supervision, termination, and intervention; and management of asset transfer. Zhang analyzed and determined the key factors for the success of PPP projects through systematic case studies, literature reviews, and expert interviews.

In short, analyzing the factors that affect the success of PPP from the perspective of specific projects can obtain specific understanding on a case-by-case basis. These factors include both macro-system factors and microtechnology factors, which greatly deepen the understanding of the factors that affect the success of PPP.

1.2.10.3 Inter-company Partnership and Strategic Alliance

Inter-company partnership refers to the partnership established by two or more companies to cooperate with each other in order to achieve specific goals. Strategic alliances are two or more enterprises that, on the premise of maintaining their independence, establish cooperative relationships based on resources and capabilities and characterized by joint implementation of projects or activities. Inter-company partnership and strategic alliance have similar characteristics with PPP, such as multicenter joint pressure, decision-making in mutual compromise, continuous conflicts, and frequent negotiations. Therefore, the theories of inter-company partnership and strategic alliance management have important implications for the study of factors affecting PPP success.

Clifton et al. studied the concept of integration alliance to manage typical PFI/PPP franchise contracts. This paper summarizes the governance process of PPP and puts forward the concept of absorbing alliance contract as a mechanism to improve the future performance of PPP projects. It is believed that the alliance is an agreement reached between all parties based on sharing risks and benefits in order to reach an agreed result. The inevitable characteristics of project alliance are as follows: in addition to the obligation that one party must fulfill, the achievement performance is expressed as collective and common obligation; all parties lose and win together; decisions made by the Project Alliance Committee must be agreed upon unanimously; the selection of all alliance project team members is based on the principle that is most beneficial to the project; it must have a strong commitment to solve problems within the alliance, rather than resorting to litigation. Alliance technology and experience have been discussed and applied many times in PPP. The concept of alliance provides hope for success, but it also brings risks. The main reason is that the alliance cannot correctly implement incentives and punishments, and the alliance parties become too close which may damage the essential responsibility and transparency of the public sector. Therefore, strategic alliances are only applicable to some specific types of projects. For long-term service contracts, although they have the potential to reduce project costs, they have not been proved.

In addition to directly absorbing the management principles and experiences of strategic alliances into PPP management, there are still a large number of studies involving the management of partnerships or strategic alliances in the general sense, which also have great implications for PPP governance. Mohr et al. systematically analyzed the characteristics of successful inter-company partnerships for the first time and established an analysis framework including partnership characteristics, communication behavior, and conflict resolution techniques and tested it with partnership data between suppliers and buyers. The results show that the main characteristics of a successful partnership are mutual commitment, coordination and trust between partners, mutual communication and participation, and conflict resolution techniques to jointly solve problems. The analysis framework is shown in Fig. 16.

On the basis of Mohr model, the relevant variables of the characteristics of successful partnership are further refined. It is believed that the partnership can continue to exist and develop only if the predetermined preconditions are met and the expectations of both parties are continuously reached in the cooperation process. A process model for the formation and development of the partnership is proposed (Fig. 17).

Lui et al. studied the relationship between the structure and process characteristics of inter-company partnership and satisfaction, especially the process. The structure is based on the characteristics of partners and prior transaction costs. It mainly studies asset specificity and reputation of partners. The process refers to the types of actions displayed in the cooperation process, mainly studying familiarity, simplicity of actions, and reciprocity of actions. Empirical analysis shows that familiarity and simplicity of action can significantly explain the variation of partner satisfaction, exceeding the two structural factors. Judge et al. specially studied the relationship between trust and control in strategic alliances and their joint effect on alliance results, mainly tested the essence of interdependence between control and trust, and developed the theoretical model as shown in Fig. 18. Empirical results show that the complementary effects of trust and control promote strategic alliance cooperation and reduce risks. In addition, it seems that the overall level of trust and control influences the alliance results by reducing actual or potential opportunistic behaviors. In addition, although trust and control are equally important, trust among partners is the most important factor linked to the success of the alliance. Hitt takes Chinese and Russian enterprises as examples to analyze the influence of institutional factors on the selection criteria of strategic partners. Institutional arrangements are mainly reflected in government legislation, the nature of property rights, and the existence of professional and commercial norms. The unstable environment will bring political and economic costs to the operation of the enterprise, so the unstable system will strongly affect the strategic decision-making of the enterprise. Institutional changes are unpredictable, while enterprises focus on short-term benefits. In short, institutional arrangements affect strategic alliance decisions.

In a word, the theory of inter-company partnership and strategic alliance has been relatively mature, especially in determining the factors for the success of partnership or strategic alliance, and some theoretical models and analysis frameworks have been developed, distinguishing structural and process factors and deeply studying the interrelation between key factors, such as the interaction of control and trust. In addition, great progress has been made in the operation of various factors, and in-depth quantitative analysis and systematic examination have been carried out. These results of the theoretical research on inter-company partnership and strategic alliance have important reference significance to the research on the key factors that determine the success of PPP.

1.2.10.4　The Perspective of Corporate Partnerships in Developing and Transition Countries

Tortajada believes that the public sector in developing countries is at a disadvantage due to the mismatch of negotiation capabilities between the public and private sectors. The development of regulatory, legal, and institutional frameworks is important to ensure the interests of the public and private sectors, especially consumers, and is illustrated by the experiences of Jordan, Morocco, and Casablanca. Abu-zeid and others have developed principles to solve institutional, legal, and practical problems in PPP according to the characteristics of the water supply industry in the Middle East and North Africa region and the application of PPP, including that the private sector must have advantages in productivity, have a clear concept and regulatory framework, implement economic incentives, effectively supervise contracts and accurately measure performance, be transparent, and reflect costs at prices, etc., and have illustrated with the experiences of Jordan, Morocco, Yemen, and Casablanca Lobina studied the impact of PPP on economy, society, politics, and environment based on the experience of private sector participation in water supply and environmental sanitation in developing countries and countries in transition, especially in Latin American countries, explained the difference between theoretical prediction and actual effect, and focused on entry restrictions, prohibition of competition, risk allocation, government regulation, responsibility and transparency, corruption, and public participation.

Queiroz found that the application level of PPP in transition countries is too low, but these countries are short of funds in infrastructure, and the gap is widening. Private sector participation in infrastructure construction and operation has great potential. Take highways as an example. The reasons for low private sector investment in transition countries include relatively low payment rates, lack of an appropriate legal framework, economic and political instability, and high risk expectations. Only by improving these aspects can private investors be more attractive. Snelson also found that governments in transition countries have limited funds. In order to strengthen infrastructure construction to promote economic development, reduce government debt, and maintain fiscal balance, PPP needs to be vigorously developed. However, many challenges are encountered. The key is to reduce risks and uncertainties, including commercial risks, macroeconomic risks, political risks, system instability, etc. Countries in transition have changed from traditional government direct provision of public services to PPP, which requires great changes and leaps in awareness, procedures, systems, market concepts, and risk-taking.

In short, relatively developed countries, developing countries, and countries in transition are facing many difficulties and challenges in developing PPP, which require continuous reform and improvement in institutional environment, governance mechanism, and government capacity. But so far, the research on PPP in developing countries and countries in transition is still at the level of experience, lacking systematic analysis, and theoretical summary.

1.2.11 The Important Role of the Government in the PPP Mode

In PPP mode, the government is one of the parties to the concession agreement; in addition, the government should provide a stable political and legal environment for the operation of the project. Therefore, compared with the traditional construction mode of public projects, the government's functions in PPP projects should be changed or adjusted. First of all, the government must examine the project more carefully; establish a reasonable, scientific, and transparent procedure suitable for the development of the project; establish the legal status of PPP; select partners through market mechanism; comprehensively evaluate the construction and operation strength of partners for infrastructure projects; select the best partners; and encourage innovation. The core of PPP mode is concession agreement, in which government departments should specify the standards of products or services, but not how to produce and provide them. According to the requirements of government departments, the private sector uses its own experience and advantages to prepare proposals for bidding, and government departments select the best. The government only stipulates the required product or service standards, leaving the private sector a lot of room for innovation. The private sector has the power to develop innovative methods, such as innovative design and innovative operation methods. On the one hand, it can meet the requirements of government departments for product or service standards; on the other hand, it can accurately calculate the cost of the entire concession agreement period and reduce the project cost as much as possible so as to win the bidding. Secondly, the government must determine a sufficiently cautious commitment mechanism to ensure the safety of assets, reduce project financing costs, and ensure the sustainability of project production or operation, so that private investment costs and operating costs can be compensated and reasonable returns can be obtained. In the PPP projects in Fazakerley Prison and Bridgend Prison in the United Kingdom, the government has promised to act as an insurer when the private sector cannot insure against commercial risks. However, when the commercial insurance company was unwilling to renew the insurance due to the prisoner riots and riots during the operation of the project, the government withdrew its original promise, which caused the private party to be unable to bear the greatly increased prison operation cost and finally led to the forced termination of the project. Third, government functions must define a clear boundary, including the boundary of assets and the boundary of supervision, such as market access supervision, price supervision, universal service supervision, etc., with efficiency supervision as the premise and guarantee of effective supervision. The realization of all this requires the government to have higher public management capability.

1.2.12 The PPP Model Establishes the Necessary Incentive Mechanism in the Concession Agreement

According to the "Report on the Origin of PPP Mode" organized by the Beijing Municipal Development and Reform Commission, only 30% of the public projects in Britain are completed on schedule, and only 27% of the projects do not exceed the budget. One of the main reasons is that the initial cost of the whole project has not been accurately calculated. Second, the government has not adopted appropriate risk management methods, and any difficulties and financial costs of projects under construction are borne by the government. Third, there is a lack of sufficient incentive mechanism. Similarly, more than 70% of PPP projects are completed on time, and no construction cost overruns are borne by government departments.

Under PPP mode, establishing corresponding incentive mechanism in the contract can well overcome the shortcomings of traditional public project construction such as overdue and overspending.

(a) Clear payment mechanism to ensure timely completion and no overspending. Under PPP mode, especially in projects where the government directly purchases services, the private sector can only obtain government payment after providing products or services required by the government. The government does not pay until the infrastructure is completed and services are provided. Therefore, the private sector will complete the project as soon as possible on schedule and get paid as early as possible. At the same time, government departments only pay for services already provided by the private sector, and do not pay for project overspending caused by the private sector. Therefore, the private sector will be very careful in calculating the cost of the entire contract period when bidding and will actively control it within a reasonable range.

(b) On the other hand, if the products or services provided by the private sector meet or exceed the standards stipulated in the original contract, a certain proportion of additional rewards can be obtained. If the products or services provided by the private sector continue to fall below the standards stipulated in the contract, then the PPP contract will be terminated, and public projects will either be taken over by the government or rebid and replaced by another private sector. This reward and punishment mechanism is a great stimulus to the private sector to provide qualified products or services. Since the private sector aims to maximize profits, when its behavior endangers this goal, the private sector has full initiative to correct the behavior deviating from the goal. The British Ministry of Finance has conducted a survey of over 500 PPP projects in operation. Survey data show that when the services provided by the projects fail to meet the standards required by the contract and are punished by payment cuts, almost all the services subsequently provided by the punished projects meet the contract requirements. Seventy-two percent of the punished projects even

provide better services than the services required by the contract after being
punished.

(c) Reasonable arrangement of capital structure, so that the interests of all participants in the private sector are consistent. The construction period and quality
control of contractors and the product quality control of equipment suppliers
are one of the major problems that have troubled the construction of traditional
public projects for a long time. It is also a problem that PPP mode can effectively solve through capital structure arrangement. Under the traditional mode,
public project construction contractors and equipment suppliers have no direct
relationship with whether public projects operate smoothly and profitably after
the project is completed and equipment is provided. However, they reduce the
quality of projects under construction and equipment within the scope that the
owners can afford to increase their profits. Contractors, equipment suppliers,
and operators have their own goals to maximize their benefits and conflict with
each other. However, in PPP mode, all parties concerned can form a joint venture or special purpose company (SPV), requiring contractors, equipment suppliers, and operators to each hold a certain share and finance through a certain
way. In this way, during the contract period, the revenue of public projects is
closely related to contractors, equipment suppliers, and operators, while the
construction quality and equipment quality of public projects directly affect the
project's operating revenue, thus making the interests of contractors, equipment
suppliers, and operators consistent. From the beginning, attention should be
paid to the performance of the project throughout its life or contract period,
overall consideration should be given, costs should be effectively saved, and
defects should be corrected in time.

(d) To determine a reasonable risk sharing mechanism, PPP projects have large
investment amount, long construction period, many uncertain factors, and great
risks to bear. Any party in the public sector or the private sector shall bear the
risks alone, which is not conducive to the successful implementation of the
project. If all the risks are borne by the private sector, once the risks arise, the
project will be difficult to implement and even end in failure, the government
will become the victim of project failure. If it is all undertaken by government
departments, it will lead to low efficiency of construction and operation in the
private sector due to soft risk constraints. Therefore, in order to ensure the success of PPP projects, the relevant risks must be reasonably shared between the
public and private sectors.

The general principle of risk sharing is each risk should be borne by the party
that can best control the occurrence of the risk, specifically, the risk should be
adapted to the bearing capacity; risk should be commensurate with control power;
risk should be coordinated with the degree of investor participation; risks should
correspond to benefits; and failure to deal with risks is equivalent to the losses
caused. Risk management is adapted to the economic impact of the project. Under

PPP mode, the government should take responsibility for the provision of public goods and services, emphasize the protection of public interests, and implement reasonable risk sharing under the concept of "win-win." At the same time, the establishment of a reasonable risk sharing mechanism can enable the private sector to reduce risks as much as possible by means of production and its own efficient operation, rather than by transferring risks to the government, thus ensuring the smooth implementation of the project and avoiding the low efficiency of construction and operation caused by soft risk constraints in the private sector. However, the government should take certain risks to ensure the profits of the private sector to a certain extent, and it can also enhance the enthusiasm of the private sector to participate in infrastructure construction.

Take Hong Kong Disneyland theme park PPP project as an example. In December 1999, the Hong Kong SAR government and Walt Disney Company signed a contract to build Hong Kong Disneyland theme park. At the same time, Hong Kong International Theme Park Co., Ltd., was established as the developer and operator of the theme park. The SAR government owns 57% of the company, and Walt Disney owns 43%. The park was opened to the public in September 2005. In this project, the Hong Kong government is responsible for land acquisition and early infrastructure construction, while the private party is responsible for the construction and operation of the park. The risk sharing is shown in Table 1.1. Its reasonable risk sharing mechanism in advance embodies the "principle of being borne by the party that can best control the occurrence of the risk," thus realizing the timely completion and opening up and the current good operating conditions. On the contrary, if the risk is concentrated too much on one party, once the risk occurs, the party who bears too much risk is easily unable to bear the risk, thus causing the failure of the whole project. The Dabo Power Plant built by Enron and the project company was the largest BOT project in India at that time. The project was built in Maharashtra State, the seat of India's largest city Mumbai, and was undertaken by engineering contractor Burkhard with equipment provided by General Electric. The prospect is very good. The project company has signed a government power concession agreement with Mabang Electric Power Bureau (state-owned), which stipulates that all electricity prices shall be settled in US dollars. The electricity generated after the completion of Dabo Power Plant is purchased by Mabang Electric Power Bureau, and the minimum amount of electricity purchased is set to ensure the normal operation of the power plant. Under certain conditions, the electricity price will be adjusted according to the power generation cost. The Malaysian government provided guarantee to the project company, and the Indian government provided counter guarantee to the guarantee provided by the Malaysian government. In this way, almost all exchange rate risks and market risks have been transferred to the Indian government. Just as the project was under construction, the Asian financial crisis broke out, and the rupee rapidly depreciated by more than 40% against the US dollar, which led to the exorbitant online electricity price of Dabo Power Plant and eventually forced the Mabang Electric Power Bureau to purchase electricity from Dabo Power Plant at a price close to twice that of other sources. When world energy prices rose in 2000, the difference rose to nearly four times. By November 2000, the Mabang Electric Power Bureau was on the verge of bankruptcy and began to refuse to pay the electricity bill for Dabo Power Plant. According to the agreement, first the state government of Malaysia and then the federal government of India temporarily allocated part of the funds and fulfilled the guarantee and counter guarantee provided. However, they could not afford the huge amount of US dollars needed to continue to fulfill their commitments and had to refuse to continue to allocate funds, resulting in the failure of the project. The failure of the project is linked to the unreasonable risk allocation structure of the project.

1.2.13 Research Status and Practice of PPP Model Abroad

The research on "public-private partnership," "public goods," and "local public goods" by foreign academic circles and some international organizations has accumulated for a long time, which is not only initiated by scholars from European and American countries but also closely related to the pace of reform and development of political, economic, and social life in these countries.

1.2.13.1 Review of Foreign Studies on Public-Private Partnerships

Research on the concept of public-private partnership. Some foreign scholars have put forward an unverifiable view on the concept of "public-private partnership." We have found from some relatively primitive documents abroad that the concept of public-private partnership originated in France in the eighteenth century. At that time, in order to solve the problems of municipal planning and urban infrastructure construction, the French government first put forward this concept. However, due to the sudden outbreak of the French Revolution in 1789, this concept failed to take action and died. At present, the public-private partnership we are talking about began in the late 1970s when Margaret Thatcher was in power in Britain. At that time, the Conservative government took the lead in proposing the "privatization" reform movement in order to get rid of the huge pressure on the British government's finances brought by the welfare state system formed after the war. The spread of this movement quickly swept the whole European and American world. With the introduction of market competition mechanism, the strong promotion of business management technology in local government management, and the popularity of PFI, the advantages of public-private partnership are highlighted in the supply of public goods for local government. Some people think that the concept pedigree of public-private partnership originated from the Australian government's new public management reform movement in the late 1980s. However, British officials and academics believe that, in 1992, Norman Lamont, then Chancellor of the Exchequer of Major's Conservative government, defined PFI (private finance initiative, the earliest term for PPP) for the first time in his "Autumn Statement 1992." He pointed out that the private financing initiative is aimed at increasing the participation of the private sector in the public service supply. On May 6, 1994, the British and French tunnel project built and operated by BOT mode lasted for 8 years, cost 10 billion pounds, and was chartered for 65 years. It was considered as a model project of PPP mode. The EU regards PPP as an important means to solve its social development projects (cornerstone), which has been implemented in many EU countries. For example, its third Poverty 3 strategy, implemented from 1989 to 1994, forms different-ent PPP projects with private capital in national and local governments, social organizations, non-governmental organizations, etc., thus causing rapid economic, political, and social changes. The European Union commissioned the "European Foundation for the Living and Working Conditions" to conduct a survey on the contribution of PPP in the implementation of anti-poverty projects from 1994 to 1997 to assess whether the PPP mechanism is an effective strategic means. The countries surveyed include Austria, Belgium, France, Finland, Germany, Greece, Ireland, Spain, and the United Kingdom. The survey conclusion is affirmative.

At present, Britain is the most important country to implement PPP mechanism. The British government regards PPP as one of the key factors in implementing its modernization strategy, covering almost all public utilities, such as transportation, prisons, water supplies, schools, and sanitation. According to statistics from the British Ministry of Finance in 2003, PPP investment in public utilities has increased rapidly since 1997, from 667 million pounds for 9 projects in 1995 to 7.6 billion pounds for 65 projects in 2002. By 2003, 451 projects had been completed, including 34 hospitals, 119 other health projects, and 239 schools of various types. In 2003–2004, PPP financing accounted for 11% of the total infrastructure investment in the same period. Transportation accounted for the highest proportion (22%), and health and education accounted for 16% in the industry classification. In terms of regional distribution, PFI projects in all areas under the jurisdiction of the British central government are more than 22, with financing of at least 400 million pounds. According to another report, from 1992 to October 2004, there were more than 380 kinds of project contracts, with a single project amount of over 15 million pounds and a total project amount of 41.5 billion pounds. Since May 1997, the British government has introduced private investment into public services through mechanisms, including 550 projects of London Underground. More than 50 countries, including Italy, Ireland, Japan, and the Netherlands, have consulted with advisers of the British Treasury on PPP. Now a PPP Industry has been formed. Canada has established the Canadian council for public-private partnership. The Brazilian government has drawn up a PPP project contract bill, and President Lula has also proposed the establishment of a trust fund to protect the interests of private investors in the PPP agreement. The main contents of the contract bill are the return period stipulated in the contract may exceed 30 years; if the contract is not well executed, compensation will be compulsorily provided to both parties; and if the contract is not executed correctly, it can be cancelled. The principle of compensation and sharing of project income is stipulated. In 2000, the British Labor Party government divided the public-private partnership (PPP) into three categories in its report "Public-Private Partnership-Government Initiative": introducing private sector ownership into state-owned enterprises or selling state-owned shares to the private sector using various property rights structures. The public sector purchases high-quality public services for a long time by signing contracts with the private sector. Selling public services to wider markets since the British privatization movement was born, there have been great disputes between the Conservative Party and the Labor Party and between the government and society. Therefore, some British scholars believe that the term "public-private partnership" (PPP) has been coined in Britain to describe the same type of things in which the private sector participates, as privatization has become politically contentious. It is not so much a radical transformation in public-private relations as a cooperative and technical act, which is a substitute for "privatization" advocated by Thatcher administration. Darrin Grimsey, a British scholar, and Mervyn K. Lewis, an Australian scholar, believe that public-private partnership is a very effective incentive-compatible contractual arrangement, in which private entities provide support for participation or infrastructure supply. Emanuel S. Savas, a world master of privatization theory and practice and an American scholar, believes that public-private partnership is a word that is less controversial than "privatization." It can define "public-private partnership" as a diversified arrangement between the government and the private sector, with the result that some or traditional public activities undertaken by the government are undertaken by the private sector. Canadian privatization theorist E.R. Yescombe believes that public-private partnership originated in the United States and must be understood from four key elements: first, it is a long-term contract between public and private sectors; second, the private sector is responsible for the design, construction, financing, and operation of public infrastructure (equipment); third, the users of the public sector or facilities will pay the private sector during the life cycle of the contract; and fourth, ownership of facilities and equipment can be retained in the public or private sectors. Similar to Jescoby's point of view are British scholars J. Barlow and others. They also believe that different types of public-private partnerships can be paid directly by the public sector (through public finance, i.e., tax revenue) or by the users of the service. Moreover, they emphasize that the private sector bears substantial financing and operational risks in such a relationship.

Investopedia, one of the world's largest international financial education websites, believes that public-private partnership is essentially a commercial relationship between a private company and a government agency aimed at completing a public service project. It is often used by government agencies to finance projects such as transportation network systems, parks, and conference centers. Using public-private partnership to finance public projects can achieve the goal of completing projects ahead of schedule.

In addition, the United States PPP National Committee, United Nations Training and Research Institute, European Commission, and Canada PPP National Committee have also defined the concept of public-private partnership according to different standards.

Research on the model of public-private partnership. The modes of public-private partnership in Sabas include outsourcing of service (outsourcing), outsourcing or leasing of operation and maintenance (O&M), cooperative organization, leasing-construction-operation (LBO), construction-transfer-operation (BTO), construction-operation-transfer (BOT), peripheral construction (wrap-around addition), purchase-construction-operation (BBO), and construction-ownership-operation (BOO). Darrin Grimsey and Mervyn K. Lewis put forward the following models: build-operate-transfer (BOT), build-own-operate (BOO), lease (lease), joint venture (JV), operation or management contract, and/or cooperative management. The United Nations Economic Commission for Europe believes that there are two main models: one is various types of joint ventures between public and private sectors and the other is the UK's PFI (private finance initiative). According to the different responsibilities and risks between the public and private sectors, the organization further subdivides the public-private partnership into the following modes: BBO, BOO, BOOT, BOT, BLOT, DBFO, finance only, O&M, DB, and operation license. The United States Agency for International Development has proposed two forms, Contract PPP, US Agency for International Development's institutionalized PPP, and the following models: DB, O&M, Turnkey operation, BOO, BOOT/concession, BBO or temporary privatization, BLOT/concession, lease-development-operation (LDO) or purchase-development-operation (BDO)/concession, design-building-financing-operation (DBFO) or WAA)/lease contract, and lease purchase.

Research on the theory of public-private partnership and privatization. There are two completely opposite views on the relationship between public-private partnership and privatization: one is that the two have different names but the same meaning; and the other is that the two refer to different things. Sabas, a master of privatization in the United States, thinks that the two are the same thing. He points out that the general trend of change in the world is to alienate the government and get closer to the society. This is the basic connotation of privatization. In other words, privatization is to rely more on non-governmental organizations and less on the government to meet public needs. In order to reduce disputes, public-private partnership has become a substitute for privatization. Public-private partnership is a diversified institutional arrangement between public and private sectors that turns various public activities traditionally undertaken by the government into private sector supply.

Competing with the above views, the two are different. British scholars Hall D., Motte R., and Davis believe that privatization (or privatization) was a political strategy of Thatcher government to rebuild the private sector in the 1980s. It is a positive and radical development strategy, and privatization covers various types of private participation from selling state-owned enterprises to outsourcing public services. The public-private partnership emphasizes the cooperation between the public and private sectors and the technical behaviors of the private sector.

Research on the theory of public-private partnership and consultative democracy. The theory of deliberative democracy is more embodied in democratic participation in the study of public-private partnership. S. Kruljac pointed out in his research on the public-private partnership and the sustainable development of the local environment that in order to realize the sustainable development of the solid waste management department to the environment, the application of the public-private partnership must be organically integrated with the consultative democracy. British scholar Tim Forsyth put forward the concept of "consultative public-private partnerships." He took the World Summit on Sustainable Development held in Johannesburg, South Africa, in 2002 as the background. It is believed that the public-private partnership is not only a short-term instrumental agreement between the government and private contractors but also a new political field involving different administrative subjects of environmental and development policies and norms. The development of new environmental governance means must fully improve public consultation in these normative fields.

Public-private partnership, local governance, and theoretical research on local government. The United Nations Economic Commission for Europe (UNECE) pointed out that the challenges faced by the government in developing effective public-private partnerships to provide public goods and services can be defined as "governance."

The government should be on guard that if they choose this kind of institutional arrangement rashly because they lack relevant knowledge on how the public-private partnership works effectively, they will risk repeating the mistakes of other countries. Therefore, if the government wants to fully benefit from the public-private partnership, it must "enable" the system, procedure, and process of the public-private partnership, and only by enabling can the government acquire the ability to govern and mature itself. UNECE has put forward the governance objectives of public-private partnerships such as fair and transparent processes, guarantee of value for money, and improvement of basic public services. The United Nations Economic and Social Commission for Asia and the Pacific (ESCAP) has proposed to ensure that "good governance" under public-private partnerships must achieve the goals of fair and transparent administrative processes, fair incentives, broad representation, and acceptable dispute resolution mechanisms.

As for the research on the relationship between public-private partnership and local government, it is generally manifested by the concern about the control of local government in the process of introducing the public-private partnership mechanism to provide public goods and services by the private sector. The United States Agency for International Development (USAID) pointed out that after the introduction of the public-private partnership mechanism, the local government cannot relax its control over the supply of public services. Moreover, in order to meet its own goals, the local government must establish a set of basic rules to regulate the public-private partnership. In all models of public-private partnerships, local governments have corresponding rights and responsibilities.

Although PPP in the modern sense has only a history of more than 10 years, foreign scholars have long discussed the mode of mutual cooperation between the public sector and the private sector. Fosle and Berger used the method of empirical analysis to study the government's approach to attract the participation of the private sector to better provide public goods and services in seven representative cities in the United States and pointed out that the appropriate public-private cooperation mode should be selected according to the natural conditions, economic structure, and political system characteristics of the city.

Brooks and others mainly discussed what roles the public sector and the private sector should play in public-private cooperation from the perspective of welfare economics and how to properly evaluate the appropriate degree of private sector participation to balance fairness and efficiency. What is more, they also made a preliminary exploration on the participation of multinational companies in public infrastructure construction in underdeveloped countries through public-private cooperation.

In 2003, US scholars M.A. Mastoid, M. Enfield, etc. compared the costs, staff salaries, and performance before and after adopting PPP mode for solid waste treatment in Libya's capital Tripoli and came to the conclusion that adopting PPP mode for solid waste treatment can use less cost but produce greater performance.

As a new way of international application in infrastructure construction, PPP mode has been widely used and has achieved success in many countries. The implementation scope of PPP projects in Britain covers transportation, environmental protection, hospitals, schools, labor, welfare, national defense, prisons, housing, government offices and community development, etc., which has had a profound impact on the political, economic, and social development in Britain. At the same time, the government has established a complete promotion system to promote the implementation of PPP projects. Later, in practice, the United States proposed to use PPP (public-private partnerships) as a special term for PFI cooperation, which was recognized by other countries, and the concept of PPP has been used up to now. Portugal started PPP mode in 1997, which was first applied to the construction of highway network. During the 10-year period to 2006, the highway mileage doubled. In addition to highways, ongoing projects include the construction and operation of hospitals and the construction of railways and urban subways. Brazil passed the "Public-Private Partnership (PPP) Model" Act in December 2004, which makes specific provisions for the state administration to implement the PPP Model Act's project bidding and signing of project contracts. According to the Ministry of Planning of Brazil, 23 highway, railway, port, and irrigation projects that have been included in the 4-year development plan from 2004 to 2007 will be the first bidding projects under PPP mode with a total investment of R $13.067 billion.

At present, PPP mode is widely used in both developed and developing countries. The private sector's investment goal is to seek projects that can repay loans and return on investment. The government's social and economic goal is to bring the greatest economic benefits to society through investment. PPP mode is the best combination of the two. Due to the great practical value of PPP, many countries such as Britain, Canada, Japan, and South Korea have set up special research organizations to discuss PPP models suitable for their own national conditions. The government has also set up special agencies to coordinate PPP projects.

1.2.14 Research Status and Practice of PPP Mode in China

The study of PPP as a formal concept in China started late, mainly the introduction of foreign PPP theories and the discussion of the feasibility of applying PPP mode in China. Most of the domestic researches on PPP financing methods stop at introducing the definition, characteristics, background, advantages, and disadvantages of PPP and other financing methods and the necessity and feasibility of using PPP financing methods in China. There is little research on the risk management of PPP mode.

1.2.14.1 The Study on Quantity and Characteristics

We searched for "public-private partnership and local public goods" and found only 64 documents and only fuzzy matching in terms of document title. These documents are mainly concentrated in 2009–2013, including 38 periodicals, 3 newspapers, 13 master's papers, and 5 doctoral papers. If we relax the scope of search, according to the essence of the concept of public-private partnership is to provide public goods and public services supply mechanism (mode or mode) search, that is, to search for "public-private partnership" as the topic of literature. PPP is generally translated as "public-private partnership." There are also translations of "public-private partnership," "public-private partnership," "public-private partnership," and "public-private partnership." We carry out literature search according to six names from English to Chinese. Generally speaking, 840 articles are entitled "PPP" in English, 66 articles are entitled "public-private partnership," 3 articles are entitled "public-private partnership," 12 articles are entitled "public-private partnership," 44 articles are entitled "public-private partnership," and 6 articles are entitled "public-private partnership" (partnership), totaling 971 articles.

According to the time distribution chart of literature retrieval (Fig. 19), we can find that the research on public-private partnership in China's academic circles and other relevant parties began roughly in 2002, with four articles in that year. From this year, the research on public-private partnership in China has shown a rapid development momentum. From 2003 to 2004, there were 9 articles and 18 articles, respectively. The first research peak was formed in 2005, with 44 articles of various types in that year. The second research peak was formed in 2006–2008, and there were 79, 88, and 94 documents in these 3 years. From 2009 to 2010, there were 126 papers and 131 papers, respectively, showing a steady growth compared with previous years. The third research peak was formed in 2011, with 160 articles of various types in that year. From 2012 to the end of 2013, there were 128 and 87 articles, showing a downward trend.

1.2.14.2 The Research Focus

The academic research on the public-private partnership involves many perspectives, and the research hotspots reflect the attention of the academic circles and all circles in our country to this problem. Through literature retrieval, we measure the hot research issues in this field in recent 10 years and find that the research on public-private partnership can be divided into the following research perspectives: infrastructure and public projects, project risks, financing modes, theoretical analysis, government management reform, introduction of foreign research, legal perspective, influencing factors and performance evaluation, concepts (features and functions), price and pricing issues, and other research. As shown in Fig. 20, we rank according to the proportion of the research literature in various fields, namely, infrastructure and public project research perspective accounts for 45%, project risk research perspective accounts for 14%, financing model research perspective accounts for 14%, theoretical analysis research perspective accounts for 8%, government management reform perspective accounts for 7%, foreign research introduction perspective accounts for 4%, legal perspective accounts for 2%, influencing factors and performance evaluation perspective account for 1.2%, concept (features and functions) perspective accounts for 0.9%, price and pricing perspective accounts for 0.9%, and other sporadic and scattered research accounts for 3%. From the above statistical data, it can be found that the research is mainly focused on project management and economics, and the research on legal issues related to public-private partnership, especially in the fields of public management and political science, is rare. The research in these three areas accounts for only 17% of the total literature, while the literature directly related to government management reform is only 7%. Research from the perspective of infrastructure, project risk, and financing mode is the mainstream of domestic public-private partnership research. Papers in these research directions are mainly concentrated in journals of architecture, economics, and management. There are the following categories of infrastructure research: first, hard economic infrastructure, including highways, airports, railways, bridge construction and maintenance, ports, rural infrastructure, land consolidation, urban renewal, etc. Second, hard social infrastructure includes hospitals, "village in city" reconstruction, drinking water safety projects, hardware construction and management of colleges and universities (such as dormitories, classrooms, swimming pools, stadiums), affordable housing, low-rent housing, public rental housing, nursing homes, and construction waste disposal. Third, soft economic infrastructure includes innovation of health science and technology mechanism and basic education and higher education mode. The fourth includes soft social infrastructure, environmental pollution control, public hospital restructuring, and home care services. The research in the field of project risk mainly includes the principle of risk sharing, the model design and risk evaluation of risk sharing (FAHP/FCE, AHP, ISM-HHM method, etc.), and risk early warning. The research and introduction of financing modes mainly include PPP, PFI, BOT, and TOT. The perspective of legal research mainly focuses on several issues such as "domestic and foreign legislative situation," "administrative law," "administrative contract," "contractual attribute," "public and private law," and "legal subject and the relationship between rights and obligations."

Among them, Jingfeng and others' "Research on Legislation and Regulation of PPP Model and Its Application in China" mainly introduced the legislative status of PPP model in Britain; briefly discussed the current legislative status of PPP model in the United States, Canada, Australia, New Zealand, Japan, South Korea, Hong Kong, China, European Commission, and other countries and regions; and put forward suggestions on the current legislative status of PPP model in China and the legislative framework to deal with the future. Tao's On the Practice of Public-Private Partnerships in China and Its Legal Framework Construction and Xiuli's "Research on Legal Issues in the Implementation of PFI/PPP in China" mainly discuss the problems in the development of PPP in China and the current legislative situation. Xiaopeng and others put forward the theoretical issues of "contract attribute," "administrative attribute," "administrative discretion," "administrative superior benefit right," and "government right to interfere regulation" of PPP contract from the perspective of administrative law in their two papers, "Study on the Attribute of PPP Contract Based on Administrative Attribute of Administrative Law" and "Study on Government right to interfere Based on Administrative Attribute of PPP Contract," by Lei and others.

According to the question of whether the administrative contract is a public law contract or a private law contract, there are mainly two papers: Zhongle and Shufen's "discrimination of Legal Issues in PPP Agreement" and (2007) "Public-Private Legal Relationship in PPP Agreement and Its System Choice." Li Xunmin, a Taiwanese scholar in China (2012), in his article "Control of Public Law Contracts-Starting from the Framework of Public-Private Partnerships," elaborated on the development of PPP in Britain, Germany, and the United States, as well as the nature of public law or private law in administrative contracts. Xiang et al.'s "The International Law Subject Status of Non-governmental International Organizations in Public-Private Partnerships" take "International Non-governmental Organizations" as the research object and analyze the international law subject status of international non-governmental organizations in transnational public-private partnerships. The research on the influencing factors of the development of public-private partnership is mainly through empirical investigation and statistical analysis to discover various unfavorable factors hindering the public-private partnership. These factors mainly include lack of local funds, inability of local government to provide guarantee to private sector, lack of unified management system of central government, unclear decision-making process, lack of experience in PPP project development and management at all levels of government, weak government PPP-related business knowledge and ability to obtain information, lack of legal system and transparent procedures to stimulate potential investors, low level of contract governance, corruption and rent-seeking behavior, weak control and supervision of PPP projects, and participation of civil society and non-governmental organizations.

1.2.14.3   Related Monographs

Since 2008, books specializing in public-private partnerships have been published gradually. Dai Jingbin's *Introduction to Public-Private Partnerships in Modern Cities* (2008) discusses the functions of public-private partnerships from the aspects of urban renewal, urban economic development, urban public sphere, urban marketing, and the role and function of urban government. Wankuan's *Governance of Public-Private Partnerships* makes an empirical analysis on the performance evaluation and influencing factors of public-private partnerships. Ying's *Path and Strategy of Private Capital Participating in Public-Private Partnership (PPP)* analyzes the environment and risks of private capital participating in PPP mode and specifically analyzes the path of private capital investing in urban infrastructure, transportation infrastructure, etc. and expounds the financing strategy from the perspective of finance. Puqu and Salamon's *A Study on Government Purchasing Public Services from Social Organizations: An Analysis of China's and Global Experiences* used a comparative analysis perspective to study the basic situation, contents, successful experiences, and lessons of failure of Chinese, British, French, and German governments purchasing public services from social organizations. Other works include Shoukui's *Study on the Mechanism and Regulatory Policy of Public-Private Partnerships in Public Projects*, Bo and Fei's *Study on Public-Private Partnerships: Based on the Construction and Operation Process of Infrastructure Projects*, Guifeng's *Study on the Supply of Public-Private Partnerships in High-Speed Railway Projects*, and Yisuo's *Study on the Promotion Law of German Public-Private Partnerships*.

Xiuhui and Shiying gave a brief introduction to the background, conceptual characteristics, advantages, and application of PPP and looked forward to the application prospect of PPP in China's public infrastructure construction.

Wang Hao made a useful discussion on the definition and classification of PPP, combined with China's actual situation and industry characteristics, and explored and studied two PPP modes that are suitable for China's rail transit projects – the former compensation mode and the latter compensation mode. He also took Beijing Metro No. 4, the first case of PPP financing scheme in China, as an example to analyze the role of the fare issue in PPP operation from the perspective of economics and public management, aiming at the most critical subway fare policy in government regulation, trying to determine the orientation of the subway fare control mode in China, and putting forward new ideas for establishing a perfect market-oriented subway fare policy suitable for China's national conditions.

Yuan Leping, Lu Mingxiang, and Li Xikun analyzed that the shortage of funds is the direct reason for the insufficient development of infrastructure in China, and diversified financing structure is the inevitable trend for the development of infrastructure industry. It is feasible in theory and practice for private capital to enter the field of infrastructure. It is necessary to alleviate the shortage of capital in infrastructure industry. It plays an important role in improving investment efficiency, promoting economic growth, and accelerating system reform. However, due to obstacles such as market access and administrative system, the entry of private capital is severely restricted. In order to resolve the barriers to the entry of private capital, measures should be formulated in terms of investment scope, financing channels and legal system to attract and encourage private capital to invest in infrastructure.

Liu Xinping and Wang Shouqing believe that PPP project financing mode has been widely used in the world, but in actual application, the public and private sectors often find it difficult to reach an agreement on the allocation of risks in PPP projects, greatly prolonging the negotiation time and increasing the transaction cost. This paper analyzes the factors that affect the risk allocation of PPP projects, puts forward a more reasonable risk allocation principle, and designs a corresponding risk allocation framework, which has a practical guiding role in guiding public and private sectors to negotiate PPP projects.

Liu Zhi elaborated the PPP mode from the angle of government management system and financing mode and analyzed the effect of applying PPP mode in public service fields in Britain and other countries. This paper makes an economic analysis on the application of PPP mode in stadiums and gymnasiums and puts forward two financing structures for legal person bidding of Beijing Olympic venues. On the basis of summarizing the bidding and financing structure of legal persons for Beijing Olympic venues, the paper puts forward the idea of implementing PPP mode in the field of public services in China.

In a word, the domestic research on PPP is still in the initial stage, and there is no systematic research on the operation mechanism of PPP and the risk sharing among all participants.

1.2.14.4 Research Practice

In 1908, when fires frequently broke out in the Imperial Palace, Empress Dowager Cixi asked Yuan Shikai what to do. Yuan Shikai said that a water supply company would be established by foreign means. Therefore, a cash register in Tianjin raised 2.7 million silver dollars to set up Beijing's first tap water company. The company's chairman wrote to the government to supervise the business. This may be the earliest PPP or similar PPP case in China. Shenzhen Shajiao B Power Plant, which was invested and built by Hong Kong and Industrial Company in 1984, is a typical case of public-private cooperation. The concession period has ended, and investors have transferred it to the local company. In recent years, governments at all levels from the central government to the local government have issued policies to encourage nonpublic capital to invest in public utilities, creating conditions for privatization of public utilities. Some local public utilities that are open or partially open absorb private capital investment. PPP mode has been tried in such basic public utilities as highways, subways, water supplies, and electric power such as Hubei Xiangjing Expressway (opened in June 2004), Hong Kong Subway, Changsha Power Plant, Beijing No. 10 Water Plant, and Veolia Chengdu No. 6 Water Plant, and some experience has been gained. The PPP model adopted in the construction and operation of Beijing Metro Line 41 is a franchise company established by social investors in cooperation with the government to participate in the construction and operation of

Line 4. The franchise company for Line 4 is owned by the Hong Kong Metro Company, a social investor. Beijing Capital Venture Group Co., Ltd., and Beijing Infrastructure Investment Co., Ltd., which are funded on behalf of the government, have a total registered capital of about 1.5 billion yuan, with Hong Kong Metro and Beijing Capital each accounting for 49% and Beijing Infrastructure Investment Co., Ltd. accounting for 2%. About 2/3 of the funds of the PPP cooperative company will use nonrecourse bank loans. PPP Company signed the Franchise Agreement with Beijing Municipal Government on February 7, 2005, in the form of public-private partnership to jointly invest, construct, and operate Beijing Metro Line 4. This is the first PPP project in China's urban rail transit construction.

The application of PPP mechanism in China is relatively late, its application scope is not wide, and its development speed needs to be accelerated. However, with the continuous, rapid, and coordinated development of China's economy, the huge demand for infrastructure will provide a broad space for private sector participation. China's public utilities need huge funds for infrastructure construction alone, and public funds alone cannot meet the demand. Of course, with the development of China's economy, especially in developed regions, there may not be a lack of construction funds for the construction of some public utilities such as health projects. However, the introduction of the new concept of public-private partnership can break the monopoly of public utilities and introduce a competitive mechanism. Financial money can be used to guide, discount, and subsidize public utilities, improve their service efficiency, and speed up the development of public utilities.

1.2.15 Problems Exposed in Public-Private Partnership Reform

PPP is a new thing emerging in the process of gradual reform in our country and lacking of institutional soil. Although there are many successful cases in the process of exploration, such as Hangzhou Bay Cross-Sea Bridge Project in Zhejiang Province, Zhuyuan No. 1 Sewage Treatment Plant Project in Shanghai City, Laibin Phase II Power Plant Project in Guangxi, Shenzhen Bus Group and Hong Kong Kowloon Bus Group Joint Venture Project, etc., there are also many failures. The most typical phenomena are:

1. Due to various considerations, the public sector requested to change or break the contract after signing the contract, which resulted in the loss of investors' interests and even the collapse of the project. Some local governments did not fully demonstrate the project when introducing capital and blindly promised too much. Once the market situation changes, or the purpose of financing is achieved, or because of government personnel changes, cross-sectoral interests are difficult to coordinate, and other reasons are the contract is required to be changed and investors are "readily available and removed." For example, the BOT project

of Tangxun Lake Sewage Treatment Plant, which was the first of its kind for Wuhan's non-state capital to enter the urban sewage treatment field, was started in 2001. After the completion of the first phase of the project, the relevant departments were unable to solve the problems such as the construction of supporting pipe networks and the collection of sewage charges. As a result, the plant was finally declared dead after being idle for 2 years. In fact, after the reform and opening up, the first batch of foreign-funded independent power developers such as the world's three major energy giants, American Mailing company, British national power company, and French Alstom company, have all closed their offices in our country and basically withdrew from our power market. These have a lot to do with the expected instability of government departments' behavior.

2. Public sector decision-making has a serious short-term orientation, eager to achieve, even beyond the legal policy authorization. Relieving short-term public finance pressure is often the fundamental starting point for government departments to adopt PPP mode. Some local governments are eager for quick success and instant benefits. They equate PPP directly with privatization of property rights. They even violate the relevant regulations on the transfer of state-owned property rights and sell the shares of public enterprises without conducting public bidding or auction. As a result, the contract is terminated or the agreement cannot be implemented at all.

3. The public interest has been ignored, the price of public goods has soared, and the public cannot bear it. The public sector and the private sector conspired to transfer risks and financial burdens to the public through short-term actions such as sharp price increases and free allocation of land, thus damaging the public's interest in obtaining services.

4. The contractual rights and responsibilities are out of balance, and the public sector has taken on excessive obligations. The contract ignores relevant policies and regulations of the state and promises a fixed return on investment, which leads to the invalidation or dissolution of the contract and increases the probability of credit risk, for example, Shanghai Dachang Waterworks Project and Shenyang No. 8 Waterworks Restructuring and Listing Project.

1.2.16   Analysis on the Causes of Practical Problems of Public-Private Partnerships

In fact, the development of PPP in our country has encountered various institutional and policy resistance, mainly from the following aspects:

1. The unequal status of cooperative subjects. Is PPP an administrative contract or a commercial contract? Or is it an economic contract or a "big civil" contract? In PPP contracts, the public sector and private enterprises are both the relative subjects in the contractual relationship, and the government is also the regulator. The

long-term model of "big government, small society" leads to the public sector often occupying an absolutely dominant position in PPP. As a result, the change of local government or personnel changes often directly affects the implementation of contracts, and the phenomenon of government breaking contracts is common. Government behavior cannot be effectively restrained, which may be the deepest root cause of market disorder.

2. Lack of social public/consumer interest appeal mechanism. The public is the ultimate user of public services, but there is no effective way to express their interests. As a result, the rights of the public/consumers to participate in decision-making are ignored, and the public interests are violated. On the other hand, it is also easy for the public to misunderstand PPP.

3. Lack and irrationality of relevant laws and regulations. At present, there is no special legislation on infrastructure or public utilities franchise in our country, and there are no clear rules on how to grant the franchise, who will grant it, the mechanism of entry and exit, dispute mechanism, etc. For example, officials from the Beijing municipal development and reform Commission said that according to the strategic concept of "new Beijing, new Olympics" and the overall goal of taking the lead in basically realizing modernization, the total investment in various infrastructure projects in Beijing will exceed 320 billion yuan from 2005 to 2008. However, the entry of social investors into Beijing's infrastructure construction and operation market is still in a situation where there is no law to follow, and relevant departments have introduced market mechanisms and strengthened market supervision without any law to follow. After the administrative license law comes into effect, franchising, as a special administrative license, can only be adopted when laws, administrative regulations, and local regulations clearly stipulate franchising. At present, with the exception of Xinjiang, which has formulated special local laws and regulations, other places basically follow the rules of ministries or local government. However, these regulations are often not perfect, and some regulations are even unreasonable. For example, at present, many central and local regulations prohibit the transfer of franchise rights, but infrastructure and public utilities franchises usually adopt project financing methods. However, when providing project financing, banks, especially international financial institutions, usually require the project company to pledge the franchise rights or the equity of the project company, resulting in difficulties in project financing.

4. The existing rules are incompatible with each other and with the international rules. The rule systems are incompatible with each other and even have obvious contradictions. For example, China's "Administrative Licensing Law" stipulates that concessions must be granted through bidding and auction, while the "Government Procurement Law" stipulates that government procurement can take the form of bidding, invitation to bid, competitive negotiation, single-source procurement inquiry, and other procurement forms recognized by the government procurement supervision and administration department of the State Council. China's current foreign exchange control system is also incompatible with general international rules.

5. The regulatory system is chaotic. The World Bank has clearly pointed out that "effective supervision" is the most critical condition to promote the correct implementation of infrastructure reform. However, in our country, there is no effective coordination mechanism and dispute resolution mechanism in the division of labor between the central and local governments, the vertical division of supervision, and the allocation of supervision resources. The supervision of PPP cooperation process is not in place, and the supervision of corporate governance of the project company is not in place. Price supervision is the core means of external supervision, but the supervision ability of price supervision department is seriously insufficient. In addition, the responsibilities among government departments and levels are not clearly defined, and the division of labor is unclear. In the process of public-private cooperation, there is a lack of predictability as to which level of department has the ability to negotiate and sign contracts on behalf of the government and whether the intention expressed by the actor is legal and effective. In addition, there are some operational problems, for example, government officials lack the basic knowledge in relevant fields and blindly make commitments that are not in line with reality. Another example is the change of market demand.

1.2.17 Establishing a Legal Framework to Promote the Healthy Development of Public-Private Partnerships

Like other reforms in our country, PPP has also taken a "practice first" road. However, practices lacking normative system guarantee are just like travelers in the dark who cannot walk fast and far. The broad development prospect of PPP in our country also illustrates the necessity of studying PPP, and the various problems exposed by PPP in our country's practice illustrate the importance and urgency of studying the legal framework of PPP, because whether any institutional arrangement created in practice can develop healthily and vigorously in the end must depend on the protection of law.

1.2.17.1 The Subject of PPP Legal Relations

The proposal and application of PPP reflects a new concept in the production mode of public facilities and services. It includes the whole process of public service provision in which the public sector selects appropriate private partners and cooperates with them to share risks and benefits. Therefore, in this sense, PPP should belong to the category of public procurement law or an extension of traditional government procurement. In the legal relationship of PPP, it mainly includes the public sector, private partners, the public/consumers, suppliers providing products or services for the construction and operation of public facilities, financing providers/loan banks, etc.

The process of PPP implementation is the interactive process of the above stake-holders. In the interrelation of these kinds of activity subjects, the most core should be the relationship between the public sector, the private sector, and the public.

The process of PPP implementation is the interactive process of the abovemen-tioned stakeholders. In the mutual relations between the subjects of these activities, the most core should be the relationship between the public sector, the private sector, and the public.

1. Public sector. PPP has changed the way public services are produced and pro-vided, but it cannot change the nature of public services themselves. The ulti-mate responsibility for ensuring the provision of qualified public services to the public remains with the public sector or the government. Therefore, in the PPP mode, the public sector bears at least three roles: public service purchaser, ser-vice quality supervisor, and service responsibility guarantor. Public service pur-chasers are relative to the private sector. Service quality supervisors or supervisors refer to the dynamic supervision of the process of public service provision. Service liability guarantors refer to the public sector's responsibility to the pub-lic once the private sector or the project company fails to perform its due respon-sibilities and obligations.

2. The private sector. From the perspective of public procurement law, the private sector is the supplier in the procurement legal relationship, providing public facilities or services that meet the requirements to the purchaser (public sector). From this point of view, the public sector is the customer of the private sector. However, unlike other contracts, although the contract is signed by the public sector, the customer of the contract, or in fact the most important customer, is not the public sector but the public service consumer.

3. The public. The public is the final consumer of products or services provided in PPP contracts and the final judge of service quality. Imagine if the goods in a supermarket are of poor quality and expensive, consumers can simply not go to the supermarket to shop, but if a bridge or drinking tap water is similar, the public has very little choice. Due to the natural publicity and long-term nature of public utilities, the public, as the ultimate consumer of services, should become an important party in PPP legal relations.

1.2.17.2 Building a Preliminary Legal Framework for Public-Private Partnerships

The United Nations Commission on International Trade Law believes that the estab-lishment of a PPP legal framework must abide by the principles of transparency, fairness, and long-term commitment. The legal framework specially established for PPP shall at least include the following contents:

1. Definition of subject qualification. The definition of subject qualification includes at least two aspects: one is to define which level of department vertically and horizontally has the qualification and ability to negotiate and sign contracts on

behalf of the government and the second is to define the qualifications of the private sector to participate in the competition for the provision of public services. Generally speaking, all potential cooperators in other fields should be treated equally without discrimination in the source and nature of capital, except that special qualification restrictions can be applied to procurement activities in specific fields (such as national defense, etc.).

2. Access system. Access system includes access scope and partner selection mechanism. The departments involved in PPP's global development have already spread all over education, police, hospitals, airports, highways, railways, subways, bridges, water, electricity, satellite manufacturing, even national defense, prisons, etc. What kind of fields are PPP's entry fields in China? Some people think that following the guidelines for foreign investment and the guidelines for foreign investment industries, private capital can be divided into four categories: encouragement, permission, restriction and prohibition according to the natural monopoly of projects and the strength of public goods, and the guidelines for private investment in public utilities can be established. The core of the partner selection mechanism is to introduce competition and make the process fair, transparent, and efficient. Bidding, invitation to bid, competitive negotiation, auction, and other forms should be included.

3. Hearing procedures. There is no need to say much about the significance and principles of the hearing procedure. PPP projects involve the long-term and vital interests of the public. The most effective public intervention is prior intervention in the hearing. How to ensure that the hearing will no longer become a decorative form, so that plastic flowers can bear fruit, is worth further discussion.

4. Supervision mechanism. At the very least, it includes the setting up of supervision institutions, the composition of supervision personnel, the price supervision mode, the service quality supervision mode, the operation supervision mode, etc. China's existing public sector regulatory agencies are divided by industry, but many PPP projects involve cross-industry and cross-sectoral issues. We can consider using the American model for reference. Each state in the United States has a regulatory body for public utilities, called the Public Utilities Management Committee (PUC), which is responsible for supervising the public utilities in the state. This horizontal integration of regulatory agencies can play a synergy effect that the original regulatory system does not have. The composition of supervisory personnel includes experts in at least three fields: law, management, and economy and finance. In most PPP forms, the private sector has the right to price services. The price of service not only determines the income of the private sector but also affects the vital interests of the public. Therefore, the price supervision mode may be the most central link in the supervision mode. According to a World Bank report, countries with strong government administrative capacity generally adopt a price cap pricing mechanism for the management of public service industries, i.e., the inflation rate within a certain period of time is taken as the upper limit, and the price rise rate of enterprises is required not to exceed this upper limit. The supervision of the operating efficiency of partners is also an

important part of the regulation of PPP. The criteria for evaluating public services generally include responsiveness, universality, quality, efficiency, and safety.

5. Protection and compensation mechanism for investors' interests. As private investors are in a weak position relative to the public sector in PPP contracts, it is necessary to properly highlight the protection mechanism of investors' interests and give private investors or project enterprises sufficient operational autonomy and the right to relief for contract breach. Judging from China's judicial practice, private investors' claims against the government are rarely supported. Government departments often use "public interest" as a privileged reason to amend or even break contracts, sinking the private sector's upfront costs. This has become the biggest concern of private investment in public utilities, thus affecting the long-term development of public utilities. At the very least, investors should be given the right to know in advance, the right to a written defend oneself, the right to bring an administrative reconsideration, and so on.

6. Exit mechanism. It is the mechanism to urge project companies that cannot provide qualified public services to terminate the provision of services. Since the operation of the PPP project company not only involves the interests of the company's shareholders, partners (public departments), and internal employees but also directly involves the public utilities and public interests, the withdrawal mechanism of the PPP project company is very sensitive. At least attention must be paid to: first, the criteria for withdrawal must be quantified; second, the exit procedure is more operable; and third, a grace period and a transition period must be given to avoid greater fluctuations.

7. Dispute handling and relief mechanism. The most frequent disputes in PPP legal relations can be roughly divided into two categories: first, disputes between the public sector and the private sector and, second, disputes between the private sector and public consumers. For the first type, please refer to "investor interest protection and compensation mechanism." In the second case, on the one hand, it can be settled through civil litigation or arbitration procedures; on the other hand, the public sector is also an important part in resolving disputes within the PPP framework. Therefore, it is very necessary for the public sector to establish relevant mechanisms to coordinate, such as complaint receiving mechanism, handling mechanism, feedback mechanism, etc.

8. Optional rights, obligations, and risk sharing modes. From a worldwide perspective, there are many modes to choose from. Judging from the increasing sequence of risks transferred to the private sector, there are service outsourcing, O&M (operation and maintenance agreement), DB (design and construction agreement), DBM (design and construction and maintenance agreement), DBO (design and construction and operation), LBO (lease and construction and operation), BLT (construction and lease transfer), BOT (construction and operation transfer), BOOT (construction and operation transfer), BBO (purchase and

construction and operation), etc. It is also necessary to establish the rights, obligations, and risk-bearing basic framework of the above model.

9. Other basic contract terms and conditions, such as duration and extension of cooperation. The duration of PPP cooperation is also one of the important factors in project design. The cooperation period is too long, which is not conducive to maintaining incentives for private cooperators and the realization of public interests. Too short, it will affect the income of private partners. Of course, there are many PPP models. It is unnecessary and unrealistic to set a uniform deadline for all models, but it is necessary to set an approximate upper limit. In addition, regarding the overall legal environment required for PPP development, the United Nations Development Programme believes that PPP needs state machinery and legal systems to ensure the legitimacy of its existence, including clear state policies and laws; effective, transparent, and independent legal frameworks at the financial and technical levels; and clear division of responsibilities among state agencies. This includes the legislative capacity of local councils to support PPP. China's reform train is running, and any instability or even change of rules or policies may bring deep or shallow impacts to PPP that has not yet established itself. In fact, institutional changes in various fields such as government transformation, investment and financing system, credit system, foreign exchange management, engineering construction, and even urban-rural integration will affect the development of PPP. Is PPP growing in oscillation or stopping in disturbance? In any case, its future will be closely related to the development direction and trend of China's reform.

1.2.18 Research Review and Enlightenment of PPP Model's Development Path in China

First of all, the government can design a new PPP model under the combination of different ownership transfer and management right control methods, combining different links such as construction, financing, operation, maintenance, and handover, which can concise and produce a PPP innovation model with Chinese characteristics. Secondly, the government should gradually reduce the inertia constraints on the selection and decision-making of several commonly used PPP modes (such as BOT, BT, TOT, etc.); expand the application scope of PPP modes in public projects on the basis of considering the public attributes, safety requirements, and operation experience of project products; and explore the operation rules of PPP modes in public projects such as outsourcing, lease-back, repurchase, franchise, and asset stripping. Finally, for a PPP project, different cooperation modes can be adopted at different stages or different links instead of a single mode.

1.2.19 *The Concept of Joint Participation of All Actors in PPP Projects*

The actors in the PPP project cooperation relationship are actually the stakeholders of the project. To study the resource allocation of PPP projects, first of all, all partners must have a common concept in the project life cycle, form a common understanding and negotiation mechanism to solve problems, and establish a supervisory role for the organizational structure of resource allocation.

1.2.19.1 Project Partner Concept

Project partnership, in essence, refers to a cooperative relationship established by two or more organizations on the basis of contracts signed by two or more parties and mutual trust to achieve win-win results for both parties, reduce risks, and improve management level. The project partner concept attaches great importance to the trust and cooperation between member partners and requires frequent communication and elimination of various frictions and doubts in project activities in order to seek common sustainable development and ultimately maximize benefits. The introduction of the concept of project partner is conducive to the formation of a relaxed and harmonious relationship between the participants, improving mutual trust and mobilizing the enthusiasm of all parties.

1.2.19.2 Project Collaboration Concept

At present, there is no unified and accurate definition of project collaboration in the field of engineering construction. The most authoritative definition is proposed by the American Institute of Architecture Industry: the management mode of project collaboration is a long-term mutual cooperation agreement reached by two or more organizations in order to achieve specific commercial interests and maximize the utilization of resources by all participants. The concept of project collaboration points out that saving resources and achieving maximum resource efficiency are the cooperative purposes of all project participants. Therefore, based on the perspective of trust and resource sharing, they reach a short-term or long-term cooperation agreement with each other and formulate relevant clauses to restrict their respective behaviors.

In view of the long-term nature and complexity of the PPP project construction process, it is very necessary to introduce a project cooperation concept that can effectively integrate resources and manage the actions of all participants and allocate and optimize resources from the overall interests of the project, so as to truly realize win-win results.

1.2.20 SuDiscussion on the Application of Public-Private Cooperation Mode in Ecological Project Financing

1.2.20.1 Macro-level

1.2.20.1.1 Give Full Play to the Functions of the Government

Successful operation projects cannot be achieved without the strong support of the government. The introduction of PPP mode in ecological construction projects is actually to break the monopoly of the government in ecological construction and to introduce market competition mechanism. Reasonable positioning of the government's role is the key factor to achieve this goal. Ecological projects have inherent particularity. It is impossible to complete large-scale ecological construction tasks by relying solely on the cash flow generated in the operation process. In the ecological construction process, the government cannot simply "retreat the country and advance the people." The government should assume the responsibility for infrastructure construction and not simply hand over the construction responsibility to the market. Ecological construction must receive sufficient support and help from the government. The government's role in ecological construction should be changed from leading role to cooperating with the private sector. The government should play the role of cooperation and supervision. The government department should protect the public interest, be responsible for the overall organization and coordination of ecological projects, and straighten out the relationship between the responsibilities and rights of all parties involved. At the same time, as the government controls ecological construction and unifies pricing power, the government department needs to give private investors transparent system expectations to ensure private profitability requirements. The long-term integration of government and enterprise in ecological construction restricts the process of the reform of the ecological investment and financing system. The government should gradually break the integration of government and enterprise and use market incentives to provide a high-quality investment and financing environment for ecological project construction, lay a good market foundation for investment and financing, and improve the economic strength of the project and the possibility of successful financing.

1.2.20.1.2 Improve the Relevant Legal System

Although there is some legal support for the application of the model in China, it is still not perfect and China lacks specific legislation. Each link of the mode operation, including design, construction, operation, and management, will involve risk definition among all parties involved, which requires clear support and guarantee of laws and regulations system to improve the operation efficiency. PPP mode is applied more and more widely in our country's infrastructure construction, and more and more problems are involved in specific applications. Therefore, it is necessary for the national legislature to establish legal norms, incorporate legislation into the national legislative process, and provide a unified and standardized code of conduct for infrastructure construction. Each local government shall formulate specific implementation rules according to the local actual situation to ensure that the project construction has laws to follow and the relations between all parties involved are stable.

Effectively solve problems such as contract disputes during project operation. At the same time, with the continuous opening of China's investment field, the application of various projects often involves international enterprises and consortia, so China's legislation should conform to international practices and conform to international standards. The monopoly advantage of the government is very obvious in the aspect of ecological construction, and private capital is in a weak position. Therefore, when applying the model, it is beneficial to form a long-term and stable partnership between the public and private sectors and give full play to the positive role of the model by using legal norms to restrict the government's behavior, limit the possibility of its discretion, and reasonably define the distribution of power between the government and the private sector.

1.2.20.1.3 Improve the Access Mechanism of the Mode

① The market access mechanism for the construction of ecological projects has been opened. The State Council's Opinions on Encouraging and Guiding the Healthy Development of Private Investment issued in January stipulates that private sector capital is allowed to enter the ecological industry. However, in the concrete implementation process, the phenomenon of "visible entry but not visible entry" and "entry but pop-up" of private sector capital should be avoided. This requires improving the investment environment of private sector capital and encouraging and guiding private sector capital to participate in the construction of ecological projects.

First of all, private capital and government capital should be given equal status and fair treatment to create a fair policy environment for private capital to participate in the construction of ecological projects. Secondly, build a financing platform for private capital to enter ecological projects and provide effective financing channels to attract private capital to participate in the construction of ecological projects. Thirdly, make public the access conditions so that all participating private capital can obtain sufficient market access information to ensure that the bidding process is fair. Finally, appropriately relax the restriction conditions for private capital to enter ecological projects, actively guide private capital to participate in the construction of ecological projects, and promote diversification of investment subjects. ② Strict regulation of franchise access qualifications. Strictly regulate access to franchising.

The government should choose partners in line with the principles of openness, fairness, and impartiality, avoid covert operations, and establish a standardized access permit system. The granting of franchise rights in our country is mainly by way of bidding. When selecting franchise objects, price cannot be taken as the only reference standard. The comprehensive abilities of the participants, such as production technology level, management experience, qualification conditions of employees, whether they have relevant project experience and performance, etc., should be fully investigated to ensure that the franchise company can successfully complete the construction and operation tasks. At the same time, the procedures for the implementation of ecological projects should be standardized as far as possible to avoid disputes arising from the differences between the parties involved in the implementation process, delay the construction progress, increase the operating costs, and achieve prior risk control.

1.2.20.2   Medium Level

1.2.20.2.1   Building a Professional PPP Advisory Service

PPP mode should be applied to all fields of public goods and services, involving disciplines including architectural design, technical consultation, financial accounting, legal affairs, financial services, and professional management. Therefore, a large number of high-quality professional consulting service enterprises are needed. In our country, the number of professional companies engaged in PPP project consulting is small, the professional standard is generally not high, and relevant industry standards have not been established. In foreign countries, some professional PPP consulting companies and related accounting, financial, and legal service agencies have accumulated rich experience and solid knowledge in the global PPP market, and some agencies have been conducting business consulting in similar fields in China for a long time.

In view of this, we should vigorously cultivate our local PPP professional consulting companies. In the process of institution building, professional PPP consultation institutions can be filed, and a PPP professional consultation service institution library can be established. In this process, we should learn from overseas institutions and appropriately introduce relevant foreign institutions when necessary, so as to accelerate the improvement of PPP practice in China. In addition, it is necessary to establish and perfect the management system and norms of PPP consulting industry. This not only requires practitioners to abide by basic professional ethics but also to be independent, impartial, and scientific in dealing with specific PPP project consultation. At the same time, PPP projects are required to be linked with consulting agencies, and relevant consulting agencies shall be held accountable for failed PPP projects.

1.2.20.2.2   Setting Up a Special Project Management Organization

The approval of PPP projects and effective supervision of the construction and operation of the projects are the keys to ensure the success of PPP projects. At present, all PPP projects implemented in our country are managed by the ministries themselves, and most of them adopt the management mode of "one thing, one discussion." This not only increases the cost of capital but also increases the cost of management. At the same time, there may be an expansion of financial risks. Therefore, we can learn from the experiences of Britain, Canada, and Australia and combine with the actual situation in our country. The Ministry of Finance can lead the establishment of a special management organization to be responsible for the approval of PPP projects and the management of PPP projects nationwide.

1.2.20.3   Microlevel

1.2.20.3.1   To Determine a Reasonable Concession Franchise Period

The project company participates in the construction and operation of the project. The investment in ecological project construction is huge, and the payback period is long. Therefore, determining a reasonable franchise period is the key to determine the profit space of the project company. When calculating the franchise period, a reasonable profit period should be calculated according to the financial model, which should not only consider the profitability requirements of the franchise company but also ensure that the government still has certain revenue and control over resources after the project is handed over.

1.2.20.3.2   Establish a Reasonable Risk Sharing Mechanism

The risks of a project cannot be borne by any single participant alone but should be effectively dispersed. Reasonable allocation of risks is a condition for the successful operation of project financing. The construction cycle of ecological projects is relatively long, the investment required is quite large, and the payback period of investment is very long. There are great uncertainties and risks. Therefore, it is necessary to establish a reasonable risk sharing mechanism to disperse risks so as to enhance the enthusiasm of private capital to participate in the construction of ecological projects. On the basis of identifying the risks of ecological construction, the following principles of risk sharing should be followed to distribute the risks of the project to different participants. The risks of any investment should correspond to the benefits. High risks must require high benefits, otherwise the investment will be diverted to other projects with higher benefits. While obtaining high benefits, the investment must bear corresponding risks: the principle of correspondence between risk and degree of participation. In the ecological project model, each participant in the project has different ways and degrees of participation. The degree of participation of the participants is positively related to the risks they bear. The higher the degree of participation of the participants, the greater the risks they bear. The risks of the project should be mainly distributed to the participants with the strongest control power. In case of uncontrollable factors, the participants should share the risks together. In general, risks should be allocated to those who are most capable, and various risks should be reasonably allocated according to the risk-bearing capacity and risk management capacity of the government and the private sector, so as to maximize the overall benefits of the project.

1.2.20.3.3 Professional Management Institutions and the Support of Complex Talents

In the traditional financing mode of ecological construction, the government is in an absolute control position. From the initiation and feasibility study of the project to the construction and operation of the project, the government is entirely responsible for the organization and subordination of the private capital. However, in the financing mode, the private capital and the government form a good partnership and are on an equal footing. They jointly participate in the construction and operation of ecological projects, and all parties involved coordinate with each other and complement each other. Therefore, the organization and coordination of specialized management agencies are needed to improve the operation level of the ecological project mode.

The application of PPP mode is a complex and systematic project, involving a wide range of areas. Professional knowledge, such as legal, technical, financial, and management knowledge, is required in fund raising, financing negotiation, project implementation, risk sharing, income distribution, etc. Therefore, the requirements for talents are relatively high. Government departments need to be familiar with the operation of the project when drafting documents and formulating relevant laws and regulations. The compound talents project has unique operation rules. Therefore, the project company needs to have professional compound talents to participate in the specific operation. Compound talents are an essential condition for the successful operation of the project. At present, China is still relatively short of talents in this field. Therefore, it is necessary to speed up the cultivation of compound talents and deepen the application of the project in China.

1.2.20.3.4 Apply Certain Compensation Mode

Ecological PPP project is a government-private cooperative construction project. The investment is huge and requires a certain amount of financial support from the government. It can be divided into pre-compensation and post-compensation according to the time of investment. The government's investment in the construction period is called pre-compensation, and the investment in the operation period is called post-compensation. The specific operation of the pre-compensation mode is to divide the ecological project into a public welfare part and a profit part. The public welfare part mainly includes civil works such as stations and tracks, which are funded by the government. The profit part mainly includes mechanical and electrical equipment such as vehicles and signals, which are completed by the project company. At the initial stage of operation after the completion of the project, the public welfare part of the assets will be leased to the project company for use; usually only a nominal lease fee will be charged or transferred to the project company for use free of charge. After the project enters the formal operation stage, the rent of the public welfare part will be adjusted according to the expected income of the project company. This will not only ensure the income of the project company but also prevent the excess income from being generated. After the end of the franchise period, all public welfare and profit-making assets will be transferred to the government for management.

The main feature of the pre-compensation mode is to divide it into two parts according to the characteristics of different parts of the ecological project. The public welfare part is invested by the government, and the one-time construction cost subsidy is provided. The profitable part uses private capital as construction capital. On the one hand, it expands the source of construction capital. On the other hand, it is driven by interests to use funds more efficiently and can realize the market-oriented income mechanism of ecological projects. Under this mode, both public and private parties can fulfill their respective responsibilities and complement each other's advantages, thus realizing Pareto improvement. However, this compensation mode still requires a large amount of construction funds provided by the government.

Post-compensation refers to the whole process in which the government and private capital jointly invest and form a project company, which is responsible for the investment, construction, and operation of ecological projects. The government needs to calculate the cost and income of the project company in advance, provide continuous subsidies for losses incurred in the operation process, ensure the market demand for future products or services of the project company, reduce the market risks faced by the project company, and ensure the profitability requirements of private capital through compensation provided by the government. After the franchise period ends, the project company also needs to hand over all the assets involved in the project to the government. The post-compensation mode is conducive to the introduction of private capital into ecological projects under monopoly operation conditions, to breaking the government monopoly, and to market-oriented financing. Through reasonable compensation and restraint mechanisms, the project can be guaranteed to achieve the unity of public welfare and operation. The difficulty in applying this model lies in the accurate estimation of costs and benefits and the influence of the established fare policy.

Chapter 2
Research Contents and Methods

2.1 Research Content

This research mainly includes three parts, theoretical research, empirical research, and countermeasure research, of which the empirical research is divided into two parts: current situation evaluation and driving mechanism analysis.

2.1.1 Theory of PPP Model

This paper summarizes the research on the influencing factors and driving mechanisms of the existing enterprises' prevention and control of land degradation and finds out the research foundation that can be referenced by them and the space for innovative research on this basis. Part of the theories of economics, management, and sociology are applied to the study of public-private partnership behavior to construct the theoretical framework of PPP model and put forward theoretical hypotheses for empirical tests.

2.1.2 The Current Situation and Evaluation of Public-Private Partnerships in the Whole Region

According to the industry classification, 144 industries were selected, which were obviously involved in the prevention and control of land degradation.

As a sample, 46 breeding enterprises, 28 resource industries, 12 ecotourism enterprises, 34 forest products processing enterprises, and 24 wood processing

© Science Press & Springer Nature Singapore Pte Ltd. 2020
Z. Meng et al., *Public Private Partnership for Desertification Control in Inner Mongolia*, https://doi.org/10.1007/978-981-13-7499-9_2

enterprises were investigated with questionnaires and data were obtained. According to this survey data, the environmental behaviors of Chinese enterprises are quantitatively evaluated and compared.

This paper divides and collects the laws, regulations, and policies on land degradation by taking the alliance city as a unit and makes theoretical analysis on the behavior of the public sector in the public-private cooperation. This paper focuses on the analysis of how the private sector will act when the public sector uses different subsidy methods to "encourage" the private sector to participate in public-private cooperation and how the public sector will continue to act or adjust its initial behavior to improve the effect of public-private cooperation when the result of the action is different from the original intention of the public sector.

2.1.3   Research on Typical Pilot Public-Private Partnerships

On the basis of investigation and understanding in the whole region, the comprehensive analysis method is used to select and establish a public-private partnership pilot, and a typical case, Yili Resources Group Co., Ltd., is selected to conduct in-depth research on the cooperation motivation, division of labor, guarantee mechanism, impact and effect of the public-private partnership, as well as existing problems and suggestions, so as to obtain its successful experience and provide reference for the operation mode of environmental cooperative governance to prevent land degradation.

2.1.4   Methods of Establishing Public-Private Partnerships

On the basis of fully referring to the successful cases of public-private partnership in China, this paper summarizes the pilot experience in Inner Mongolia region and puts forward a suitable method for public-private partnership in Inner Mongolia, which will be popularized and applied.

2.2 Research Methods

This paper mainly adopts the method of combining theoretical analysis and empirical research, trying to deeply discuss the application of public-private partnership model in the prevention and control of land degradation in China. The following research methods are mainly adopted:

1. Literature research method. On the basis of collecting and reading relevant works, periodicals, cases, and other documents at home and abroad, we fully absorb and draw lessons from existing theoretical achievements and practical experience, refine relevant theories and viewpoints, accumulate useful materials and materials, and lay a good foundation for special research.
2. Case study method. Preventing land degradation under PPP mode involves the interests of government, private enterprises, and farmers. Different financing structures have different decision-making mechanisms and will produce different social effects. By using the method of example analysis, this paper analyzes the examples of PPP project implementation and analyzes the property right structure of public-private cooperation, the formation mechanism of cooperation, risk sharing, and project repurchase decision.
3. Deduction method: The basic theories and methods in management, economics, public management, and other fields are comprehensively applied to combine the PPP mode with the reality of our country, to discuss the application of PPP mode in ecological construction of our country, and to propose a complete set of technical systems.

2.3   Technology Roadmap (Fig. 2.1)

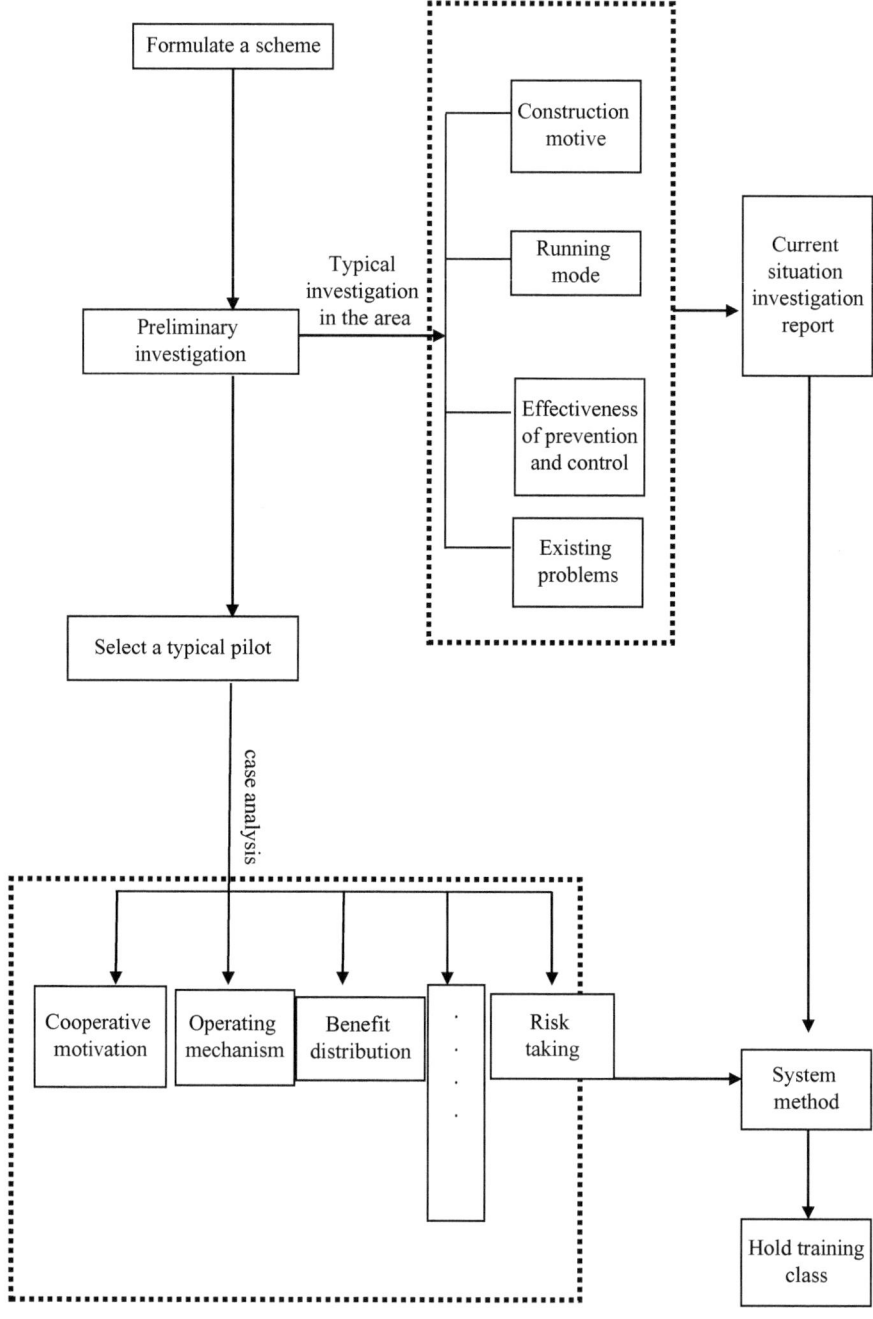

Fig. 2.1   Technical route

Chapter 3
Land Degradation

3.1 Nature and Degree of Land Degradation

3.1.1 Definition of Land Degradation

From an ecological point of view, land degradation refers to the deterioration of plant growth conditions and the decline of land productivity. From the perspective of system theory, land degradation is the result of the combination of human and natural factors. In essence, the basic connotation and process of land degradation are reflected in soil degradation, including physical, chemical, and biological degradation. In recent years, "soil degradation" has been widely used to replace land degradation in the world. However, just because the land is a natural complex of rocks, landforms, climate, hydrology, and biology, its structure and function are far beyond the scope of soil, and it is not comprehensive to replace land degradation with soil degradation. In addition, land degradation is a very complex and comprehensive dynamic process, which contains a strong concept of time. The so-called degradation and non-degradation should be understood by comparing the quality and quantity of land in different periods, such as deserts, Gobi, snow, and some rocky desert areas. For a long period of time, it is the same in quality and quantity, so it cannot be incorporated into degraded land. Specifically, land degradation is a significant decline in land quality (quantity) during the comparison period.

According to the actual situation in Inner Mongolia Autonomous Region, land degradation should be defined in the United Nations Convention to Combat Desertification (UNCCD) discussed by the Inner Mongolia Strategic Action Plan Working Group, that is, land degradation refers to the reduction or even loss of bio-economic productivity and complexity of rain-fed lands, grasslands, pastures, forests, and woodlands in arid, semiarid, and dry subhumid areas due to the combined effects of land use and other influencing factors.

© Science Press & Springer Nature Singapore Pte Ltd. 2020
Z. Meng et al., *Public Private Partnership for Desertification Control in Inner Mongolia*, https://doi.org/10.1007/978-981-13-7499-9_3

3.1.2   Types of Land Degradation in the Inner Mongolia Autonomous Region

At present, there is no unified plan for classification of land degradation types at home and abroad, but most researchers classify mainly according to the causes and consequences of land degradation. Inner Mongolia Autonomous Region has a vast territory and complicated types of land degradation. This classification mainly uses a two-level system to describe the types of land degradation in Inner Mongolia Autonomous Region. The first level is divided by the dominant factors leading to land degradation, and the second level is divided by vegetation type (ecosystem).

The main types of land degradation occurring in Inner Mongolia Autonomous region are as follows.

3.1.2.1   Wind Erosion

The phenomenon that soil materials are transported off the ground due to the action of wind and the friction and wear of particles in the airflow flowing to the ground.

3.1.2.2   Water Erosion

Phenomena caused by precipitation during soil movement and precipitation.

3.1.2.3   Salinization

The phenomenon that the accumulation of soluble salt in soil that is harmful to plants.

3.1.2.4   Freezing and Thawing

Accumulation of soluble salts in soil is harmful to plants.

3.1.2.5   Other Impact Factors

It is mainly caused by mineral resource development, urban construction, and traffic road construction. In addition, pesticides, fertilizers, agricultural films, industrial wastes, and domestic wastes cause soil pollution in agricultural production.

According to the classification of ecosystem in Inner Mongolia Autonomous Region, the types of ecosystem degradation are forest, grassland, desert, wetland, and farmland ecosystem.

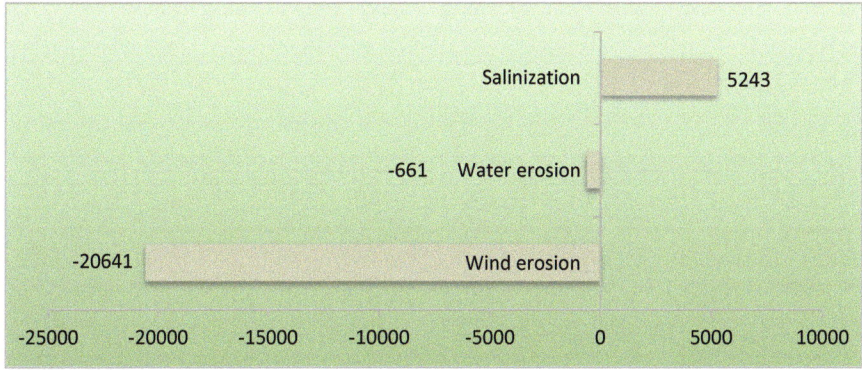

Fig. 3.1 Dynamic changes in desertification from 1999 to 2004

3.1.3   Historical Trends and Past Assessments

The area of land degradation in Inner Mongolia Autonomous Region is relatively large, and the statistical data of relevant departments are incomplete and overlap with each other. Therefore, we will use major industry data for historical trends and past assessments, such as wind erosion mainly used in forestry sector data and water erosion mainly used in water sector data (Fig. 3.1).

3.1.3.1   Wind Erosion

According to the Inner Mongolia Desertification Bulletin issued by the Inner Mongolia Forestry Department in July 2005, by 2004, the area of wind erosion and desertification land had decreased by 20,641 km^2 compared with 1999.

3.1.3.2   Water Erosion

According to the statistical report of remote sensing monitoring of soil erosion in Inner Mongolia Autonomous Region in 1995, the area of light soil erosion in the region is 150,219.00 km^2, accounting for 12.52% of the total area. According to the remote sensing survey of soil erosion in 2000, the area of soil erosion in the whole region is 147068.06 km^2, accounting for 12.43% of the total area, accounting for 18.00% of the area of wind, water, and ice. In short, soil erosion decreased from 1995 to 2000. According to the report of the Ministry of Forestry, the area of desertified land decreased by 661 km^2 from 1999 to 2004. However, according to the final data of China's soil and water conservation strategy report, the area of soil erosion in Inner Mongolia Autonomous Region in 1990 and 2000 was 158,100km^2 and 150,200 km^2, respectively, and the area of soil erosion decreased by 7882 km^2 in 10 years.

Table 3.1 The change status of land salinization in Inner Mongolia Autonomous Region

Index	Up to 1986	Up to 1986
Salinized land area/hm²	3,160,000	2,100,000
Salinized land area accounted for the total land area (%)	2.78	1.8
Secondary salinized land area/hm²	466666.67	5,4700*
Secondary salinized land accounted for the total land area (%)	0.040	0.051
Annual expansion rate of secondary salinized land/(hm²/a)		617.99

Source: Investigation and Evaluation of National Ecological status: North China Volume. Edited by the State Environmental Protection Administration
*It shows that the correlation coefficient of variables is significant at 0.01 level, and the correlation coefficient of variables is significant at 0.05 level (double-tailed test)

3.1.3.3   Salinization

Soil salinization in Inner Mongolia has zoning. In addition, human activities have also caused some secondary salinization. It is widely distributed in the whole region and increases gradually from east to west.

According to the report of the Ministry of Forestry, the salinized land has increased by 5243 km² compared with 1999. According to statistics from the 1980s, the salinization area of the whole region is about 3.16 million hectares, of which the salinization area of grassland is about 540,000 ha, and the secondary salinization land caused by unreasonable irrigation is about 470,000 ha. According to statistics of the 1990s, the salinization area of grassland was 450,000 ha, 900,000 ha less than that of the 1980s, the salinization area of farmland was 1.15 million hectares, and the salinization area was 3.94 million hectares (the salinization area of soil was 2 million hectares). In addition, the total salinization area and the total salinization area of farmland increased by 78 ha and 680,000 ha, respectively, compared with the 1980s (Table 3.1).

3.1.3.4   Freezing and Thawing

According to the statistical report of freeze-thaw monitoring by remote sensing in Inner Mongolia Autonomous Region in 1995, the freeze-thaw area of the whole region is 47.699 km², accounting for 3.98% of the total area. According to the remote sensing freeze-thaw survey in 2000, the freeze-thaw area of the whole lightly above area is 47712.71 km², accounting for 4.03% of the total area, accounting for 5.84% of the area of wind, water, and ice. In short, the freeze-thaw area in Inner Mongolia has increased slightly, mainly due to climate. Freezing and thawing are mainly distributed in Daxing'anling in northeast China. The freezing and thawing distribution in Daxing'anling in permafrost region is approximately the same as that in cold temperate coniferous forest in mountain area.

3.1.3.5   Other Impact Factors

It is mainly caused by mineral resources development, urban and traffic road construction, soil pollution caused by pesticides, fertilizers and agricultural films, and industrial wastes generated by industrial enterprises. In addition, ecological damage and mineral resources development are mainly reflected in the production, treatment, and more compensation of gangue and tailings. Secondary geological disasters in the development of mining enterprises in Inner Mongolia are mainly reflected in the severity of soil erosion and the increase of desertification area. In short, the area and degree of land degradation have increased year by year in recent years.

3.1.4   Status

According to the classification method of land degradation adopted in this strategic plan, there is currently no specific department to conduct a comprehensive survey of land degradation, so there is no data set available. According to the functions of the departments concerned with the prevention and control of land degradation in Inner Mongolia Autonomous Region and the authorities in different fields, the main functional departments have integrated the data describing the land degradation in Inner Mongolia Autonomous Region.

3.1.4.1   Wind Erosion

According to the desertification bulletin issued by the forestry department of the autonomous region in July 2005, the climate is dry and windy, and land desertification is dominated by wind erosion. The area is 563,500 km², accounting for 47.62% of the total area of the autonomous region and 90.52% of the total area of land desertification.

According to the remote sensing survey of soil erosion in 2000, the wind erosion area of the whole region is 623,300.47 km², accounting for 52.60% of the total area and 76.16% of the area (slightly above) eroded by wind, water, and frost.

3.1.4.2   Water Erosion

According to the desertification bulletin issued by the forestry department of the autonomous region in July 2005, the desertification of soil erosion in Inner Mongolia Autonomous Region is mainly distributed in the central and eastern hilly areas, an area of 27,500 km², accounting for 4.43% of the total land desertification area.

According to the remote sensing survey of soil erosion in 2000, the area of soil erosion in the whole region is 147068.06 km², accounting for 12.43% of the total area: wind, water, and frozen erosion areas (slightly eroded or above).

3.1.4.3 Salinization

According to the desertification bulletin issued by the forestry department of the autonomous region in July 2005, the salinized land area in Inner Mongolia Autonomous Region is relatively small, mainly concentrated in the low-lying areas between Hetao plain, Tumen river plain, and several sand hills, with an area of 31,300 km², accounting for 5.05% of the total land desertification area.

According to the results of the second soil survey, the area of saline soil in the whole region is 86.381 million mu, accounting for 4.9% of the total area, of which 23.88 million mu is saline soil and 62.501 million mu is saline soil.

3.1.4.4 Freezing and Thawing

According to the remote sensing survey of soil erosion in 2000, the freeze-thaw area of the whole region is 47712.71 km², accounting for 5.84% of the total area and 76.16% of the area (slightly above) eroded by wind, water, and frost.

3.1.4.5 Other Impact Factors

According to the remote sensing survey of soil erosion in 1995 and 2000, the erosion area is 329.85 km², and the water and wind erosion area is 79,900 km². The table below summarizes the basic situation of land degradation caused by other factors (Tables 3.2, 3.3, 3.4, and 3.5).

3.1.5   Land Degradation Types and Relative Importance of Ecosystems

3.1.5.1 Forest Ecosystem

The total area of forest land in Inner Mongolia Autonomous Region has remained stable for nearly 10 years, but the ecological function of forest has obviously weakened. As far as forests are concerned, the forest landscape dominated by coniferous forests has gradually been replaced by broad-leaved forests and mixed coniferous and broad-leaved forests or has become a blank. As broad-leaved forest and mixed coniferous and broad-leaved forest are mostly secondary forests, a large proportion of plantation forests exist after coniferous forests are cut down, and the forest quality and ecological function decline. Because broad-leaved forest and mixed coniferous and broad-leaved forest are mostly secondary forests, but after coniferous forest is cut down, there is a large proportion of artificial forests, and the forest quality and ecological function decline. The most prominent manifestations are the depletion of forest recoverable resources, the decline of water storage function, soil erosion, and loss of biodiversity.

Table 3.2 Ecological damage and restoration caused by industrial and mining development in the Inner Mongolia Autonomous Region

Index	As of 1986	As of 1999
Type of mining enterprise	Clay, schist, sand, gold, copper, iron, lead-zinc ore, magnetite, limestone, coal, graphite, kaolin, gypsum, sand, rare earth, other nonmetallic ores, etc. *	Coal (1722), petroleum (2), geothermal (1), iron (122), manganese (2), complex iron (2), copper (21), lead (25), zinc (19), molybdenum (11), tungsten (3), tin (4), Shajin (4), rock gold (90), silver (6), vein quartz (20), fluorite (46), limestone (2), dolomite (20), sandstone (15), refractory (8), pyrite (4), mirabilite (22), natural alkali (40), alkali limestone (50), mud-carbon lake (13), natural brine (1), mineral water (23), graphite (14), talc (16), mica (1), gypsum (21)
Number of mining enterprises	1827*	4342
Total area of destroyed land in mineral exploitation years/hm^2	153,988,6*	179,124.59
Of which: woodland/hm^2	57,247.1*	62,976.01*
Grassland/hm^2	73,673.8*	83,904.34*
Farmland/hm^2	11,254.5*	12,634.42*
Other types/hm^2	11,093*	19,609.82*
Soil pollution caused by mineral exploitation/hm^2	7027.7*	10,204.59*
Damaged land restoration area in mineral exploitation/hm^2	3448.2*	112.28
Pollution land area of mineral exploitation/hm^2	4066.8*	16.91
Rate of ecological reconstruction of mineral resources development (%)	4.88*	0.07
Ecological damage area caused by engineering construction/hm^2	130,078.3*	158,516.5*
Of which: the ecological damage area caused by traffic construction/hm^2	43,749.5*	57,541.9*

(continued)

Table 3.2 (continued)

Index	As of 1986	As of 1999
Ecological damage area caused by water conservancy and hydropower projects/hm²	76,458.6*	88,346.8*
Ecological damage area caused by water conservancy and hydropower projects/hm²	2200*	4953*
Ecological damage area caused by other projects/hm²	7530.2*	7554.1

Source: National ecological survey and evaluation: North China volume. Edited by the State Environmental Protection Administration
*It shows that the correlation coefficient of variables is significant at 0.01 level, and the correlation coefficient of variables is significant at 0.05 level (double-tailed test)

Table 3.3 Pesticide use and pollution in Inner Mongolia Autonomous Region

Index	As of 1986	As of 1999
Pesticide application with annual total/(kg/a)	3,150,700*	4,000,000
Average application with unit area/ (kg/hm²)	11.233*	0.68
Proportion of pesticide application area to total cultivated land area (%)	24.17*	70
Key counties (districts, cities)	Da Banner, Hang Jin Banner*	Da Banner, Hang Jin Banner*, Corleone's front Banner, Zahlet's Banner

Source: National ecological survey and evaluation: North China Volume. Edited by the State Environmental Protection Administration
*It shows that the correlation coefficient of variables is significant at 0.01 level, and the correlation coefficient of variables is significant at 0.05 level (double-tailed test)

3.1.5.2 Grassland Ecosystem

Due to human and natural factors, the grassland ecological environment in Inner Mongolia is relatively serious, which is manifested in the following aspects: grassland degradation, grassland desertification, grassland gravel, grassland soil erosion, grassland salinization, sand dune activation, grassland degradation, and sandstorm. In addition, the quality and productivity of grassland resources have deteriorated significantly, and rare and endangered plants are on the verge of extinction.

According to grassland types, desert grassland (west of Xilinguole, north of Wulanchabu, north of Bayannur, and central Ordos) has the highest area of degraded grassland, followed by typical grassland. The proportion of degraded grassland in

Table 3.4   Fertilizer use in Inner Mongolia Autonomous Region

Index	As of 1986*	As of 1999
Fertilizer application with annual total/(kg/a) (folding purity)	282,248,112	2,900,000,000
Average application with unit area/(kg/hm²) (folding purity)	734.281 25	300
Proportion of fertilizer application area to total cultivated land area (%)	31.12	90
Key counties (districts, cities)	Tuzuo Banner, Tuo County, Saihan District, Da Banner, Hangjin Banner, Wulanchabu City Flag County	Bayannur City

Source: Investigation and Evaluation of National Ecological status: North China Volume. Edited by the State Environmental Protection Administration
*It shows that the correlation coefficient of variables is significant at 0.01 level, and the correlation coefficient of variables is significant at 0.05 level (double-tailed test)

Table 3.5   Pollution of agricultural film in the Inner Mongolia Autonomous Region

Index	As of 1986*	As of 1999
Agricultural film area/hm²	5748.9	723,000
Average usage/(kg/hm²)	221.7	57
Proportion of area used to total cultivated land (%)	1.95	12.4
Average residual rate (%)	4.2–100	75
Key counties (districts, cities)	Toxian County, Left Banner, Saihan District, Da Banner, Hangjin Banner	Tok County, Tuzuo Banner, Saihan District, Da Banner, Hangjin Banner, Keyonfront Banner, Zarate Banner

Source: Investigation and Evaluation of National Ecological status: North China Volume. Edited by the State Environmental Protection Administration
*It shows that the correlation coefficient of variables is significant at 0.01 level, and the correlation coefficient of variables is significant at 0.05 level (double-tailed test)

forest grassland is the lowest. This reflects that the harsh environment is also the basic limiting factor for degradation.

3.1.5.3   Desert Ecosystem

The deserts of Inner Mongolia are distributed in the central and western regions. The main ecological problems of desert ecosystem are grassland degradation, soil coarsening, and desert salinization. These three types of degradation are the direct influencing factors of grassland degradation. Substrate degradation is more difficult to recover than grassland degradation, resulting in the decline of ecological protection function.

3.1.5.4   Wetland Ecosystem

There are many kinds of wetlands in Inner Mongolia, but their distribution is very uneven. Wetlands are widely distributed in humid and semi-humid forests and forest grasslands. However, wetlands are sparsely distributed in arid and semiarid grasslands, desert grasslands, and desert areas. Wetland landscape covers an area of 900,000 ha, mainly distributed in the central and western regions of Inner Mongolia. Due to natural and man-made reasons such as drought, reclamation, and upstream closure, the wetland ecosystem in Inner Mongolia has the following problems: the decrease of wetland water resources, the reduction of wetland area, the decline of wetland flood control and water storage function, the increase of downstream flood disasters, and the decline of biodiversity. The main problems existing in rivers are low utilization efficiency of water resources, serious uneven distribution of water resources, water pollution, harmful to biodiversity and high flux change rate, which should cause soil erosion. Wetland, as the kidney of nature, is also facing problems such as reclamation and unreasonable development of tourism resources, resulting in the decline of ecological functions such as flood control and water storage. In addition to natural factors such as climate, sandstorms that continue to occur in recent years are also the main causes of wetland destruction and loss in desert areas.

3.1.5.5   Agriculture Ecosystem

According to the remote sensing survey of ecological environment in Inner Mongolia at the end of the twentieth century, sand, soil erosion, and salinization in the farmland landscape reached 3.927 million hectares, accounting for 24% of the farmland landscape area. Among them, soil erosion farmland area was the largest, with 2.1035 million hectares, mainly distributed in the southern Ordos and Chifeng. Salinized farmland has a landscape area of 1,152,200 ha, mainly distributed in Hetao area. In addition, the desertification area is 645,000 ha, mainly distributed in the northern part of Yinshan (Table 3.6).

3.1.6   Sensitive Areas for Land Degradation and Distribution in Ecosystems

According to the characteristics of Inner Mongolia's ecological environment and future social and economic development, early warning is given to several major ecological systems in Inner Mongolia: agricultural, grassland, and forest ecological environment, and large-scale exploitation of mineral resources.

According to the results of the remote sensing survey of the ecological environment in Inner Mongolia at the end of the twentieth century, the ecological environment in Inner Mongolia is mainly distributed in the desert landscape in the central and western parts of Inner Mongolia, with an area of 28.95 million hectares,

Table 3.6 Relative importance of land degradation in different ecosystems

A. Environmental importance

Degeneration types	Forest ecosystem	Grassland ecosystem	Desert ecosystem	Wetland ecosystem	Agriculture ecosystem
Wind erosion	Uncorrelated	Extremely important	Important	Uncorrelated	Very important
Water erosion	Very important	Important	Important	Uncorrelated	Extremely important
Salinization	Important	Important	Important	Very important	Important
Freeze-thaw	Important	Important	Important	Important	Important
Others	Important	Important	Important	Important	Important

B. Economic importance

Degeneration types	Forest ecosystem	Grassland ecosystem	Desert ecosystem	Wetland ecosystem	Agriculture ecosystem
Wind erosion	Very important	Extremely important	Very important	Uncorrelated	Extremely important
Water erosion	Important	Important	Important	Important	Extremely important
Salinization	Important	Important	Very important	Important	Extremely important
Freeze-thaw	Very important	Important	Important	Important	Very important
Others	Very important	Important	Important	Important	Extremely important

C. Social importance

Degeneration types	Forest ecosystem	Grassland ecosystem	Desert ecosystem	Wetland ecosystem	Agriculture ecosystem
Wind erosion	Very important	Extremely important	Very important	Uncorrelated	Very important
Water erosion	Very important	Important	Uncorrelated	Uncorrelated	Extremely important
Salinization	Very important	Important	Important	Extremely important	Very important
Freeze-thaw	Important	Important	Important	Important	Important
Others	Important	Important	Important	Important	Important

D. Cultural and historical importance

Degeneration types	Forest ecosystem	Grassland ecosystem	Desert ecosystem	Wetland ecosystem	Agriculture ecosystem
Wind erosion	Very important	Extremely important	Very important	Uncorrelated	Extremely important
Water erosion	Very important	Very important	Very important	Very important	Extremely important
Salinization	Extremely important	Extremely important	Very important	Very important	Very important
Freeze-thaw	Extremely important	Important	Important	Important	Important
Others	Very important	Important	Important	Very important	Very important

Note:

Extremely important		Very important		Important		Uncorrelated	

accounting for 58% of the total area. Secondly, it is mainly distributed in grassland landscape, with an area of 15.75 million hectares, accounting for 32% of the total area, most of which is induced by human activities, causing serious damage to the ecological environment. In the farmland landscape, the worst area is 3.9 million hectares, mainly distributed in the northern part of Yinshan and loess hilly region. In addition, there are 900,000 ha of wetland landscape, mainly distributed in the central and western regions of Inner Mongolia.

According to the ecological function zoning report, the region is divided into 8 primary ecological zones, 19 secondary ecological zones, and 75 ecological function zones (Table 3.7).

According to the Report on Strategic Environmental Impact Assessment of the 11th Five-Year Plan for National Economic and Social Development in Inner Mongolia Autonomous Region, the main ecological function protection areas in Inner Mongolia Autonomous Region include Alxa Desert Oasis, Alxa Haloxylon Forest, Helan Mountain National Forest, agropastoral ecotone at the northern foot of Yinshan Mountain, water and soil erosion area in Yellow River Basin, Nenjiang River, Upper Reaches of Xiliaohe River, Daxing'anling Forest, Western Daxing'anling Forest Grassland, Mu Us Sandy Land, Hunshandake Sand Land, Horqin Sand Land, and Hulunbeier Sand Land, with nature protection as the main goal. Ecologically sensitive areas include 162 (2005) national, district, and county-level nature reserves to be built in Inner Mongolia, such as natural secondary forests in Daxing'anling, Yanshan, Yinshan, Helan, and upper reaches of Liaohe River.

3.1.7 Distribution and Characteristics of Potential Geological Hazards

On the basis of understanding the regional geological environment conditions and the main controlling factors of geological disasters, combined with earthquake trend prediction and comprehensive analysis of meteorological data, the main areas of geological disaster prevention and control in recent years are as follows:

1. From Daqingshan to the low mountains and hills at the southern foot of Wula Mountain, such as Xinghe County, Zhuozi County, Tumen River Left Banner and Tumen River Right Banner in Baotou City, Shiguai District, Bayannur City, Ural Front Banner, etc., debris flow, collapse, and landslide are prone to occur in gully and slope cutting area before quarrying.
2. Low hills in the east and south of Daxing'anling, such as Yakeshi in Hulunbuir City, Saranton, Yakeshi in Juquan county of Xingan League, Qianqi of Coyou, Zarote Banner in Tongliao City, Wuniute Banner in Chifeng City, Songshan District, and Kalarqin Banner, are prone to debris flow, collapse, and landslide.
3. Debris flow and landslide are easy to occur in low mountain and hilly areas of Ordos Plateau, such as Wuhai City, Dongsheng District of Ordos City, Zhungeer Banner, Dalat Banner, Hangzhou Banner, etc.
4. Collapse and landslide hazards are easily induced in the cutting section of Wuda-Bayanhaote Expressway in the low mountain area of the western foot of Helan Mountain.
5. The main mines in this area are Sichuan, Yuanbaoshan, Pingzhuang, Hulun Buir, Dayan, Zalenore, Wuhai, Dongsheng, Wulanchabu, Zhungeer, and Bainaimiao

Table 3.7 Main ecological characteristics of the primary ecological areas in Inner Mongolia Autonomous Region

Ecosystem areas	Sensitivity of ecological environment	Main problems of ecological environment
I. Ecological zone of coniferous forest in the cold temperate zone of northern Daxing'an mountains	Extremely sensitive to biodiversity, sensitive to soil erosion	Biodiversity is reduced, the original ecosystem is threatened, vegetation is destroyed, and water source protection function is weakened
II. Deciduous broad-leaved forest and forest steppe ecological area in the central and southern part of Daxing'an mountains	It is very sensitive to soil erosion and biodiversity and very sensitive to grassland soil erosion and desertification	Deforestation and the formation of large secondary forests have seriously damaged natural vegetation and reduced the water conservation function of primitive forests, resulting in grassland degradation and desertification, serious soil erosion, and low soil fertility in farmland
III. Ecological region of grassland in the central and eastern part of Inner Mongolian Plateau	It is highly sensitive to desertification and soil erosion, biodiversity and secondary salinization of soil	Excessive reclamation has resulted in serious soil erosion, general decline of grassland productivity, serious degradation in some areas, activation of sand dunes due to overgrazing of sand vegetation, deterioration of grassland quality, and threat to biodiversity
IV. Ecological region of desert steppe in Inner Mongolian Plateau	It is extremely sensitive to soil desertification, soil erosion, and secondary salinization	Grassland degradation is serious, soil erosion, desertification, gravel, soil erosion, secondary salinization, sparse vegetation, reduction of high-quality forage grass and serious overloading are the main sources of sandstorms in North China
V. Steppe desert ecological area in central Inner Mongolian Plateau	Extremely sensitive to biodiversity and soil erosion	Grassland degradation, soil erosion, land desertification and biodiversity reduction, soil and water conservation capacity decline; grassland degradation is serious; desert activation is an important source of sandstorms in northern China
VI. The mountain desert ecological region of the central-western-north mountain of Inner Mongolian Plateau	Sensitive to land desertification and biodiversity	Ecological environment deteriorated, *Haloxylon ammodendron* forest, *Populus euphratica* forest, and *Tamarix chinensis* forest died in large areas, oasis degenerated and shrunk, and desert grassland degenerated
VII. Ecological region of Loess Plateau	Sensitive to soil erosion and land desertification	Soil erosion, land desertification, vegetation degradation
VII. Northeast plain with agricultural ecological region	It is highly sensitive to soil erosion, biodiversity, and desertification and extremely sensitive to soil salinization	Natural vegetation has been severely damaged, land degradation is serious, soil erosion is serious, soil fertility is declining, biodiversity is declining, soil and water conservation capacity is declining, and soil is secondary salinization

copper mines in Baotou. Due to the continuous extension of underground mined-out areas, these mines are prone to new ground fissures and ground subsidence.
6. The five deserts of Badain Jilin, Tengri, Ulanbuh, Baijin Windsor, and Kubuqi, as well as the five sandy lands of Mauusau, Hunshandak, Muqin, Horqin, and Hulunbeier in Uzur and their surrounding areas from west to east in the autonomous region, are extremely vulnerable to the influence of movable sand dunes.

In addition, some areas and zones with strong human engineering activities, such as highway, railway, mining development, industrial park, and hilly and mountainous slope cutting, are prone to geological disasters such as collapse and landslide. This is also an important area with potential geological disasters.

3.2 Reasons of Land Degradation

3.2.1 Natural Factors

Natural factors of land degradation include climatic factors, matrix factors, and vegetation factors. Climate factors mainly include rainfall, strong winds, and freezing and thawing. The main factors of matrix are geology, topography, and surface soil type. Vegetation factors include vegetation coverage, vegetation type, and vegetation composition.

Among the climatic factors, the time synchronization of drought and gale seasons and the concentration of rainfall are the important causes of land degradation. In addition, its influencing factors include frost-free period, annual average number of windy days, humidity, and other factors. If frost-free period is long and annual precipitation is high, it can play a positive role in protecting water and soil. However, more wind, higher evaporation, and lower air humidity will accelerate land degradation. In high-altitude and high-latitude areas, seasonal (winter/summer) or 1-day (day/night) fluctuation of freezing point and increase of soil moisture caused by rainfall and melting of ice and snow will lead to swelling of frozen soil particles on the surface of soft soil, which is the main reason for freeze-thaw erosion and also the important factor for wind erosion and water erosion. The main factors of matrix are loose sandy sediment, thick weathering crust, loose surface composition, and low content of soil organic matter. Take the content of soil organic matter as an example. If the content of soil organic matter is high, the land production capacity is relatively good and the yield is large, the overload problem can be relatively alleviated, and the further degradation of the land can also be alleviated. If the vegetation coverage area is large, the surface roughness and stability will increase, which will weaken the wind speed and reduce the surface wind erosion and soil erosion, and the surface water environment will be relatively improved, thus promoting the increase of primary productivity (plant productivity), easing the contradiction between man and land, and curbing the further development of land degradation.

3.2.2 Direct Causes

3.2.2.1 Driving of Economic Benefits

The yield of pasture is far lower than that of cultivation. Opening a factory is easy to lose money and close down. In contrast, people often choose low-input and high-output economic benefit-driven planting and production methods. In order to alleviate the difficulties in production and workers' lives, some forest industry enterprises have opened up land to increase their income and engage in "wage fields."

3.2.2.2 Reasons for Land Reclamation Policies

For example, some regions have handled the policy formulation of "combining agriculture and animal husbandry and guiding agriculture and animal husbandry," "comprehensive forestry management," "mature land," "development of land resources," "opening up fire barriers," and "opening up fire prevention roads" well in practice. These policies have played a positive role. However, in some areas, improper handling has caused conflicts between agriculture and animal husbandry, destroyed the ecological environment, and caused ecological deterioration. In some cases, the concepts of land reclamation are called "cultivated land," "fodder land," "improved grassland," and "mature forest land," which are often intertwined and lead to management confusion.

3.2.2.3 Inadequate Laws and Regulations, Land Reclamation, Land Degradation, and Unclear Subject of Land Management Functions

Incomplete laws lead to indiscriminate land reclamation, making it difficult to investigate illegal land reclamation. Under such circumstances, when various factors lead to "reclamation fever," supervision and management are often inadequate and passive.

Governments at all levels and land management departments have done a lot of work to curb indiscriminate reclamation. However, land reclamation among relevant departments leads to inconsistent understanding of law enforcement subjects of land degradation and unclear formulation of laws and regulations, which leads to difficulties in law enforcement. For example, when some land bureaus investigate and deal with the problem of random reclamation, the court ruled that the land bureau is not the subject of law enforcement, so the land bureau lost the case.

3.2.2.4 Management Reasons

At present, there are still some problems in the management of land resources in various industries. First of all, it is difficult to manage land uniformly. In particular, the three departments of agriculture, forestry, and grassland each have their own management. Animal husbandry issues "grassland certificate" and agriculture issues "contract certificate." In many fields, certificates are repeatedly issued and overlapped. The forestry department will divide the forest edge and grassland into suitable forest land and issue "forest right certificate." The agriculture and animal husbandry department will divide the forest edge and grassland into grassland and grassland certificate, which is extremely unfavorable to stop excessive reclamation. Although the Land Administration Law controls the abuse of land and the protection of cultivated land to some extent, the protection of grassland and forest land is far from enough. In addition, some areas are "small government enterprises" and "enterprises in front of the government." Some state-owned farms, enterprises, and institutions do not have corresponding land management agencies with the local government. The local government has lost control over the land management work of these units. These units are also often responsible for land degradation caused by illegal activities.

3.2.2.5 Other Factors

The destruction of vegetation by human beings also includes the following aspects: the exploitation and utilization of mineral resources cause a large amount of destruction of surface vegetation. In addition, industrial production is the biggest invisible killer among many factors leading to land desertification. The main impacts are industrial pollution and waste of water resources, which destroy the normal growth conditions of vegetation, lead to the destruction of vegetation structure, and directly lead to land desertification.

3.2.3 Underlying Reasons

Judging from the consequences of land degradation, land degradation is mainly a kind of harm to social economy. Therefore, the potential causes of land degradation can be discussed from both social and economic aspects.

3.2.3.1 Social Factors

1. Policies and politics are potential drivers of land degradation, and the lack of scientific management accelerates the process of land degradation.

2. Policy instability: under the one-sided "food-oriented" policy, the "great leap forward" and "cultural revolution" caused serious environmental disasters in many parts of China from the late 1950s to the early 1970s. Regardless of the objective laws of the agricultural ecological environment, a large-scale reclamation campaign was launched. Most grasslands are restricted by climatic conditions, which is conducive to the development of animal husbandry. Animal husbandry or the combination of agriculture and animal husbandry should be developed. However, the excessive concentration of power makes local governments have no right to choose suitable land use methods and are forced to destroy forests and open up wasteland. Reclamation of grasslands has seriously damaged the agricultural ecological environment.

   Poor management: due to unclear "right to use, raise, and protect" grasslands, grasslands in many areas have been overstocked for a long time, resulting in serious degradation of grasslands. The survival rate of tree planting is low, and excessive logging and illegal logging are common. The preservation rate of trees in some areas of the three northern shelterbelts is only 40%, and no trees are planted every year. After the country contracted out the land to the farmers, there was no quality requirement for the operators to maintain the land, which led to the decline of land fertility. However, the farmers had no responsibility and objectively included the short-term behavior and predatory management of the land users. The imperfect management system cannot effectively control the abuse or destruction of desertified land, resulting in the continuous expansion of desertified areas.

3. The population is large and the quality is low: the population expansion has increased the demand for food and land, and the population growth is not commensurate with the national economic growth, forcing farmers to turn their eyes to grassland and have to destroy the grassland, triggering famine to produce more food. In many areas, the low cultural quality of the population; outdated concepts, conservative and closed; and not accepting new ideas and new technologies restrict the rational development and utilization of land resources. The shortage of talents also makes it difficult to popularize advanced production technology and management experience, and the intensification of agriculture and animal husbandry is low.

3.2.3.2 Economic Factors

Economic backwardness is a powerful driving force for the deterioration of the ecological environment. Low economic level will inevitably lead to the deterioration of the ecological environment. Poverty forces people to exploit and utilize land resources in a predatory and unreasonable way, overgrazing, felling, and collecting herbs, all of which are economic behaviors that lead to land degradation. Due to the imbalance between economic benefits and ecological benefits, the agroecological economic system has been severely damaged, forming a vicious circle of "worse farming and more farming."

3.3   Consequences of Land Degradation

3.3.1   Environmental Impacts in the Past and Present

Soil and vegetation: The direct result of soil erosion is the loss of soil that makes the topsoil fertile. The Yellow River Basin alone loses 180 million tons per year, while the Liaohe River Basin loses 120 million tons per year. Annual soil loss reached 300 million tons, equivalent to 3 million tons of organic matter. Desertification will lead to deterioration of soil structure, loss of soil water and nutrients, and aggravation of land poverty. Land degradation leads to a decline in soil productivity, the main consequence of which is a decrease in vegetation. In turn, this leads to a decrease in vegetation coverage and a decrease in organic fertilizer content. All of these have harmful effects on the soil, which in turn increases the risk of further soil erosion, thus exacerbating degradation.

Water resources: The reduction or loss of surface vegetation has had a negative impact on the hydrological conditions in many waters in the region. The main manifestations are the reduction of surface runoff, shrinkage of lakes, salinization of lake water, and lowering of groundwater level.

Due to the destruction of surface vegetation, large areas of dry and fine soil are exposed and eroded under the influence of strong winds and bad farming practices, resulting in the continuous decline of air quality in Inner Mongolia, especially in spring. Due to wind erosion, the dust content in the atmosphere is very high, and land degradation is the cause of frequent occurrence of natural disasters such as sandstorms.

Chemical pollution: Most farmers have increased the use of pesticides and fertilizers in pursuit of grain output per unit of land, sometimes unbalanced or excessive. This may lead to secondary pollution, rainfall, and infiltration of agricultural irrigation, which is also one of the factors causing the increase of water pollution.

Biodiversity: Land degradation may lead to the loss of biological species in the ecosystem and even the disappearance of many endemic species. This will also reduce the number of valuable active gene banks. For local communities, they have lost plants and animals that are important to their culture and economy. In addition, although there are a large number of biological species in Inner Mongolia Autonomous Region, many animal and plant species are threatened by land degradation.

3.3.2   Economic Losses in the Past and Present

Economic losses caused by land degradation are huge, but there is no systematic method to calculate them. In Inner Mongolia Autonomous Region, the main forms of land use are farmland and grassland degradation, wind erosion, water erosion, and other types of land degradation, resulting in different degrees of farmland and grassland degradation. Therefore, economic losses can be calculated according to the two major types of land use. The direct economic loss of degraded grassland is the reduction of its output benefit, such as the reduction of grassland yield, indirect economic loss, reduction of soil nutrient content, and reduction of livestock capacity. According to the calculation method of economic loss of desertification, economic loss caused by serious degradation is calculated as 100%, moderate degradation is calculated as 50%, and slight degradation is calculated as 25%. According to the national framework, indirect economic losses caused by land degradation are calculated at 2–8 times the direct economic losses and five times the median. The data quoted in the table below are those of 2004, so GDP was calculated in 2004.

As can be seen from the table, the annual economic losses caused by land degradation in Inner Mongolia Autonomous Region are staggering. In 2004 alone, the direct economic losses caused by the degradation of cultivated land and grassland amounted to 18.729 billion yuan, while the indirect economic losses amounted to 93.645 billion yuan, totaling 112.374 billion yuan, accounting for 41.43% of the annual GDP (Table 3.8).

3.3.3 Impacts of Social Welfare in the Past and Present

1. Poverty: There is a close link between land degradation and local poverty. The poorest banners are usually located in ecologically fragile and rapidly deteriorating areas. Despite the government's efforts to change the poverty situation in the country in the past few decades, millions of people still live below the poverty line. According to the preliminary estimation of the statistics department of the autonomous region, by the end of 2004, there were still nearly 1 million people in extreme poverty in the rural and pastoral areas of the whole region, and the low-income population exceeded 1.1 million. According to a survey conducted by the poverty alleviation office of the autonomous region, 450,000 people live in extreme poverty, and 1.17 million low-income people do not have enough food and clothing. The two projects need to support a total of 1.62 million poor people. Poor families cannot give up unsustainable resources development to protect resources. As land degradation, food production, livestock, and forest products decrease, poor families cannot meet their basic needs. The interaction between poverty and land degradation has led to increased demand for welfare, food relief, and poverty reduction by national and local governments.

2. Food security: Because land degradation has a significant impact on food production, livestock reproduction, and forest product production, which in turn reduces the ability of individual households to obtain food. This may be because they cannot produce enough food on their own land or because their work on other lands is not enough to buy back what they need.

3. Health: Health problems are caused by poverty, declining food production, and diseases caused by water quality. As a result, wind erosion and the increase of

Table 3.8 Estimation of economic losses of land degradation in Inner Mongolia Autonomous Region

Degenerated types	Distribution area	Degenerated degree	Degenerated area/ha	Economic loss calculation method	Direct economic loss/100 million	Indirect economic loss/100 million	Total	占GDP比例/%(2004年为2712.08亿元)
Grassland degradation	Eastern region (including Hulunbeir City, XingUNTA, Tongliao City, Chifeng City, and Xilinguole League)	Mild	20338954.99	The direct economic loss of degraded grassland is calculated by the annual output value of unit ha grassland in different regions. The eastern region is 76 yuan, the central and eastern part 57 yuan, and the central and western part 38 yuan. Different degrees of degradation multiplied by the corresponding coefficient (mild degradation times 0.25, moderate degradation times 0.5, severe degradation by 1, the same below). According to the concept of national framework, indirect economic loss is calculated as 2–8 times of direct economic loss and 5 times of median value	3.86	19.32	23.19	0.85
		Moderate	3908809.59		1.49	7.43	8.91	0.33
		Severe	2159435.42		1.64	8.21	9.85	0.36
	Central and eastern part (including Ordos, Ulaanchab, Hohhot, and Baotou cities)	Mild	4138223.82		0.59	2.95	3.54	0.13
		Moderate	1736878.95		0.50	2.48	2.97	0.11
		Severe	4295097.23		2.45	12.24	14.69	0.54

Central and Western China (including Wuhai City, Bayannur City, and Alashan League)	Mild	1807282.89		0.17	0.86	1.03	0.04
	Moderate	5764307.54		1.10	5.48	6.57	0.24
	Severe	2639682.90		1.00	5.02	6.02	0.22
Farmland degradation	12 cities in the region	Mild	3099932.24	Direct economic losses are calculated in terms of corn output per ha of farmland. According to the statistical yearbook of the autonomous region, the output value of maize in 2004 was 6790 yuan, and the indirect economic loss was also calculated as 5 times	52.62	263.11	315.73	11.64
		Moderate	1146704.26		38.93	194.65	233.58	8.61
		Severe	1221563.51		82.94	414.72	497.66	18.35
Total					187.29	936.45	**1123.74**	**41.43**

Note: The calculation method used in this part is mainly based on the calculation method of desertification loss

dust content in the atmosphere have increased the severity of health problems, affecting every family and society.

Personal safety and security: Land degradation increases the destructiveness of natural disasters, for example, floods, landslides, and mudslides not only directly threaten people's lives but also may erupt in the near future, which makes people anxious and reduces their sense of security.

4. Forced relocation: Due to land degradation, families or entire communities may be forced to move out of their homes. Due to sand dune movement, many areas have to carry out ecological migration again and again. After soil erosion, fertility decline, and land degradation, families or entire communities are forced to migrate. When there is no land to plant crops, they have no choice but to give up farm work and work in towns (Table 3.9).

3.3.4 Importance Assessment

The impact of land degradation is scattered. It is mainly reflected in the following aspects: the expansion of degradation scope, the deepening of degradation degree, and the expansion of influence scope. Sandstorms caused by soil erosion and desertification have spread to other regions. Sandstorm in Northeast China is related to land degradation in Inner Mongolia, which is the main dust source of sandstorms. In the spring of 2000, dozens of experts from the Institute of Atmospheric Physics of the Chinese Academy of Sciences, the Beijing Institute of Meteorology, the Institute of Geography and Resources of the Chinese Academy of Sciences, and the Institute of remote Sensing applications of the Chinese Academy of Sciences also made in-depth studies on the causes of dust weather in North China. It is pointed out that more than 80% of sandstorms are loam and sandy soil. Dust comes from about 250,000 km^2 of degraded grassland and abandoned farmland in Inner Mongolia and Hebei Province. In Siwangzi Banner and Wuchuan County in Inner Mongolia, scientists found that the soil on the ground was blown up by the wind, leaving behind sand that is difficult to cultivate.

Sandstorms cause traffic disruption, casualties of people and animals, building damage, communication disruption, and environmental quality deterioration in large- and medium-sized cities through sand burial, wind erosion, strong wind attacks, and air pollution. This has seriously affected people's normal production and life and poses a great threat to the construction of an ecological, civilized, and harmonious society.

3.3.5 Impacts of Land Degradation

The consequences of land degradation affect all levels of Inner Mongolia Autonomous Region and have different degrees of impact on various industries. However, due to the differences in industrial fields and geographical locations, the

Table 3.9 Socioeconomic problems caused by ecological disasters and ecological destruction in Inner Mongolia Autonomous Region

Types of ecological damage		Number of victims (including death toll)	Economic loss/10,000	Crop yield decreased (%)
Flood	1950s	677,810(101)	43,135.8	10–30
	1960s	785,339(173)	31,467.6	9–27
	1970s	5,484,892(98)	586,066.5	6–30
	1980s	5,870,247(278)	1,308,392.4	7–40
	1990s	5,329,088(134)	1,306,864.2	9–60
Drought	1950s	1,634,258	42,181.6	7–65
	1960s	2,644,711(23)	92,816	2–61
	1970s	10,414,061	301,566	6–60
	1980s	13,108,004	623,060.5	6–60
	1990s	10,745,749(9)	460,465.1	8–60
Sandstorm	1950s	1,464,695	1184.4	2–20
	1960s	2,009,125	1439.2	3–21
	1970s	2,412,317	32,718.3	2–23
	1980s	2,957,675(11)	44,966.1	2–38
	1990s	1,679,380	113,610.7	2.5–39
Water and soil erosion	1980s	432,220	60,736.0	4–22
	1990s	332,590	26,970.0	5–30
Sand (stone) desert	1980s	306,060	19,256.0	3–33
	1990s	347,100	23,988.6	4–45
Salinization	1980s	218,000	14,122.3	1.2–18
	1990s	259,000	13,642.3	1.1–23
Rivers and streams broken	1980s	10.3	5368.5	1.2
	1990s	13.6	7432.3	1.6
Others	1950s	4.3	480.0	0.2
	1960s	3.8	963.2	1.9
	1970s	260,005.2	7459.0	0.5
	1980s	300,004.6	8938.0	0.1
	1990s	640,004.2	11,693.0	0.3

Source: National ecological survey and evaluation: North China Volume. Edited by the State Environmental Protection Administration

impact is also different. The most direct ones are rural farmers, communities, local governments, and autonomous region governments. The impact on farmers and herdsmen's families in rural and pastoral areas mainly refers to the loss of income caused by the decline of land productivity, which leads to the decline of per capita income, the further deterioration of the production and living environment caused by the increase of population, and the continuous spread of poverty. It will also affect communities and local and autonomous governments, but land degradation will lead to slow or even retrogression of local economic growth, which can be used to improve and protect the ecological environment compared with the impact of farmers and herdsmen. Investment in social welfare and poverty alleviation has been increasing.

3.4   Forecast of Land Degradation Trend

3.4.1   Characteristics and Extent of Land Degradation

The types of land degradation in Inner Mongolia include wind erosion, water erosion, salinization, freezing and thawing, etc.:

Wind erosion: Inner Mongolia Autonomous Region Desertification Status Bulletin shows that wind erosion and water erosion areas have been effectively controlled.

Salinization: According to statistics from forestry and water conservancy departments, the salinization area in Inner Mongolia Autonomous Region is expanding.

Freezing and thawing: The abovementioned is also a growing trend. In addition, land degradation caused by other factors has increased significantly.

According to recent development, Inner Mongolia Autonomous Region has made remarkable achievements in preventing and controlling land degradation in recent years. The focus of people's attention is wind erosion. Two types of water erosion have been effectively prevented, but at the same time, some of this degradation types have increased. Besides freezing and thawing, other degradation types are closely related to human activities. In other words, in the future process of controlling land degradation, it is necessary to strengthen the human management of these degraded areas.

The information of land degradation in Inner Mongolia is limited, and its development trend is restricted by many factors. It is difficult to predict what will happen in the future. However, according to historical trends, current situation, and the strength of ecological environment control, we can make the following prediction: as investment and governance efforts continue to consolidate the existing governance results, wind erosion and water erosion will slow down the two main types of degradation. If governance ceases, existing achievements will be lost and the field of desertification will continue to expand. However, land degradation caused by salinization and other factors will continue in the short term. In recent years, with global warming, freeze-thaw degradation is likely to intensify. Therefore, the original ecological control project should continue. In addition to other types of degradation, land degradation types caused by salinization and other factors should be mainly controlled, and freeze-thaw work should be mainly controlled. If the investment does not increase, the current economic losses will increase and continue. If we do not solve the problem of land degradation in poor counties, it will become more and more difficult to solve the problem of rural poverty. By increasing financial and human resources investment in land degradation control programs, we can benefit economically, socially, and ecologically.

3.4.2 Forecast of the Impact of Land Degradation on Economic Losses and Social Welfare

For a long time, Inner Mongolia Autonomous Region has implemented an investment mechanism of state investment, local matching, and labor input. In recent years, with the implementation of the national key ecological projects, the national efforts to prevent and control land degradation in Inner Mongolia have increased significantly. With the development of social economy, Inner Mongolia's supporting capacity has also been strengthened, and the absorption of social capital and foreign capital has gradually increased. However, compared with the task and difficulty of controlling degraded land in Inner Mongolia, the overall investment is still insufficient and cannot meet the needs of preventing and controlling land degradation in Inner Mongolia. The current development trend is likely to be maintained if the existing methods for preventing and controlling land degradation are continued. However, if we reduce the intensity of investment management, the result will only get worse and worse, even to an irreversible point. If we increase investment management and optimize management mode, it is possible to reduce economic losses and promote social development. In addition, Inner Mongolia has a large area of degraded land and a fast economic development. Therefore, it is necessary to improve the existing governance model and speed up the governance process to surpass the speed of destruction.

Chapter 4
Efforts to Control Land Degradation

4.1 Development Plans

4.1.1 Provisions of the State and the Autonomous Region on the Prevention and Control of Land Degradation in the Ninth Five-Year Plan and the Tenth Five-Year Plan

During the Ninth Five-Year Plan for National Economic and Social Development, the state demanded "to achieve basic self-sufficiency in food based on domestic resources" with a self-sufficiency rate of not less than 95% and a net import volume of not more than 5% of domestic consumption. At the same time, the reserved cultivated land resources should be developed and reclaimed for more than 300,000 aha years to make up for the occupation of cultivated land in the same period. In addition, it is also very important to maintain a stable cultivated area and a long-term stable grain crop planting area of about 110 million hectares. The state plan for increasing grain production in Inner Mongolia is 3.5 million tons. Therefore, during the "Ninth Five-Year Plan" period, Inner Mongolia put forward the planning target of transforming 28 million mu in central China and building 4.8 million mu of "wasteland suitable for farming." The White Paper on China's Grain Problem also proposes the use of non-grain resources, believing that China's grasslands and mountain areas are rich in resources and have great potential for development. If most of them are built on artificial grassland, the level of grassland animal husbandry will be improved. China's mountainous areas account for 70% of the total land area and have good conditions for the development of woody food. The prospect of increasing woody food is also very broad. Therefore, the scale of grassland construction and forage grass base construction in Inner Mongolia also expanded greatly during the "Ninth Five-Year Plan."

© Science Press & Springer Nature Singapore Pte Ltd. 2020
Z. Meng et al., *Public Private Partnership for Desertification Control in Inner Mongolia*, https://doi.org/10.1007/978-981-13-7499-9_4

In the 10th National Economic and Social Development Plan, Inner Mongolia took ecological construction as the foundation and starting point according to the requirements of the western development. The trend of ecological deterioration should be basically curbed. We must adhere to the principle of "overall planning, step-by-step implementation, highlighting the key points, first making things easier and then making things difficult." We must focus on the five key areas of the Yellow River, the upper and middle reaches of the Yellow River, sand and alkali control areas, the Inner Mongolia sand source management area around Beijing and Tianjin, the natural forest conservation zone in Daxing'anling, the Hulunbeier Xilinguole grassland protection management area, and the Alashan natural enclosure management area. The eight major projects such as grassland ecological construction and protection, natural forest resource protection, returning farmland to forest and grass, sand prevention and control, ecological construction in key counties, "three north" shelterbelts, green passage, and soil and water conservation have been fully implemented. We will strengthen the construction of nature reserves. One hundred million mu of desertified land will be controlled within 5 years. We should improve and implement various policies for ecological construction, adhere to the principle of "whoever creates who owns," and fully mobilize the enthusiasm of the masses and the whole society to recommend afforestation. In areas where returning farmland to forests and grasslands, the livelihood of farmers and herdsmen must be effectively solved to achieve the goal of "returning farmland to forests and grasslands". Vigorously promote advanced and applicable technologies to provide scientific and technological support for ecological construction. According to local conditions, promote "advance, retreat, and return" in agricultural areas. In pastoral areas, we should "protect, transform, and develop";, successfully implement "sealing, flying, and building" in desert areas; strengthen management and protection; and improve survival rate and preservation rate. We will conscientiously strengthen law enforcement, protect the ecological environment according to law, and resolutely prohibit illegal activities such as deforestation. The improper development during the Ninth Five-Year Plan has been corrected (Table 4.1).

4.1.2   Provisions on Combating Land Degradation in the Eleventh Five-Year Plan of the State and the Autonomous Region

The overall guiding principle of the 11th Five-Year Plan is to change the mode of economic growth, realize sustainable development, and promote economic growth, from mainly relying on increasing resource input to mainly relying on improving resource utilization efficiency.

The main objective of social and economic development is to basically curb the deterioration of the ecological environment, to initially ease the restriction of land

Table 4.1 Provisions on combating land degradation in the Ninth Five-Year Plan and Tenth Five-Year Plan

	Ninth Five-Year Plan	Tenth Five-Year Plan
Overall requirement	To maintain the sustained, rapid, and healthy development of the national economy	Ecological environment and water conservancy construction are listed as the focus of infrastructure construction
	We will improve the region's overall economic strength, sustainable development capability, and national quality	Strengthening the basic position of agriculture and animal husbandry
		Implementing the sustainable development strategy
Struggle objective	Forest cover increased to 15.6%	Over 17% with forest cover
Major mission	Transforming 28 million mu of medium- and low-yield field	Strengthening ecological environment and infrastructure construction
	New 4.8 million mu of wasteland suitable for agriculture	Focus on building major ecological and environmental projects
	5 million mu of afforestation per year	The trend of curbing the deterioration of ecological environment
	The area of nature reserve is 7.2 km^2	Improving environmental conditions for economic and social development
Policy and measures	Strengthen the construction of basic dry farmland and control of soil and water conservation in small watersheds and develop water-saving ecological agriculture	The focus is on the control areas of soil erosion, sandstorm, and saline alkali in the middle and upper reaches of the Yellow River, Beijing and Tianjin sandstorm source control areas, natural forests in Daxing'anling, conservation zone, and Hulun buir and Xilinguole grassland protection and control areas. Five key areas of Alashan natural fence management area. The area of soil erosion has increased by 25 million hectares, and the grassland area has increased by 16.5 million hectares through the "three reforms"
	Take comprehensive measures to guide poverty alleviation and development immigrants from all over the country in a planned way	
	On the basis of moderately reclaiming wasteland suitable for farming and accelerating the conversion of farmland to forests and pastures, the planting area will be stabilized and the grain yield per unit area, and total grain output will be steadily increased	

(continued)

Table 4.1 (continued)

	Ninth Five-Year Plan	Tenth Five-Year Plan
	Focusing on the task of increasing grain production by 3.5 million tons in Inner Mongolia according to the national plan, we will continue to focus on comprehensive agricultural development and the construction of commodity grain bases. Select areas with good conditions and great potential for increasing production to implement intensive development	Eight major projects, including grassland ecological construction and protection, natural forest resource protection, returning farmland to forests and grasslands, desertification control, ecological construction in key counties, "three north" shelterbelts, green corridors, and soil and water conservation, have been fully implemented. Strengthen the construction of nature reserves. Five years of desertification control land 100 million mu
	The livestock industry should strengthen the construction of water-centered pasture and forage base construction	Adhere to "who creates who has," arouse the enthusiasm of the whole society afforestation recommendation
	Forestry adheres to the principle that the state, the collective, and the individual attach equal importance to construction, management, closure, and protection and strive to realize the unification of economic benefits, ecological benefits, and social benefits. We should give full play to the important role of forestry in ecological construction and strengthen vegetation restoration in desertification areas such as the Yellow River basin and the West Liaohe River Basin. To do a good job in the construction of "three north" shelterbelts, plain greening projects, and sand control projects, vigorously develop shelterbelts, timber forests and economic forests, strengthen the construction of forestry projects in mountainous and sandstorm areas, and carry out mountain closure, sand closure, afforestation and grass planting in a planned way	Promote advanced and applicable technology and provide scientific and technological support

(continued)

Table 4.1   (continued)

Ninth Five-Year Plan	Tenth Five-Year Plan
Strengthen environmental, ecological, and resource protection. Adhere to the simultaneous planning, implementation, and development of economic construction, urban and rural construction, and environmental construction. Strengthen comprehensive land management, ecological construction, and environmental protection. The focus is on the energy development zone bordering Shaanxi and Inner Mongolia, the comprehensive land control project in Alashan region, and the wind erosion and desertification control project at the northern foot of Yinshan Mountain	According to local conditions, promote "advance, retreat, retreat" in the agricultural field. In pastoral areas, we should "protect, transform, and develop"; successfully implement "sealing, flying, and building" in desert areas; strengthen management and protection; and improve survival rate and preservation rate
	Protect the ecological environment in accordance with the law and resolutely prohibiting deforestation, wasteland, and other illegal activities.
	Nierji, Nichol and other large-scale water control projects. In arid and water-deficient areas, small-scale rainwater collection projects, such as water cellars, dry wells, and reservoirs, should be built according to local conditions to do a good job of artificial rainfall
	Water-saving irrigation technology should be widely popularized in dry farming areas, and a number of demonstration projects of water-saving irrigation artificial forage base should be constructed
	To solve the problem of drinking water for 3.4 million people and 9.9 million livestock in arid and water-deficient areas and areas with high arsenic and fluorine content

and resources on economic and social development, and to basically establish a disaster prevention and mitigation system in key geological disaster prevention and control areas. During the 11th Five-Year Plan period, the forest coverage rate reached 20%, the effective utilization rate of agricultural irrigation water increased to 0.5%, the coverage rate of basic geographic information of 15,000 land reached more than 95%, and the annual renewal rate reached 20%.

Policy Measures We will promote the development of the western region, consolidate and develop the achievements of returning farmland to forests, continue to promote ecological projects such as returning farmland to grass and protecting natural forests, strengthen vegetation protection, strengthen the control of desertification and rocky desertification, and strengthen the prevention and control of water pollution in key areas. Strengthen the control of soil erosion and desertification in black soil in western northeast China. We will support the development of ethnic minority areas and border areas, protect natural ecology, and improve infrastructure conditions. To promote the construction of the main functional areas, the future population distribution, economic distribution, land use, and urbanization pattern should be considered according to the carrying capacity of resources, current development density, and development potential. Land space is divided into four types: optimized development, key development, restricted development, and forbidden development. Adjust and perfect regional policies and performance evaluation according to the main function orientation, standardize the spatial development order, and form a reasonable spatial development structure.

According to the country's local restrictions on regional functional orientation and development direction, the following are involved in Inner Mongolia:

- Non-protective logging is prohibited in the forest ecological function area of Daxing'anling. Planting trees, saving water, and protecting wild animals.
- The desertification control area of Hulunbuir grassland in Inner Mongolia forbids excessive reclamation, improper firewood collection, overgrazing, and returning grazing to grassland to prevent grassland degradation and desertification.

According to the degree of desertification, Inner Mongolia Horqin Desertification Control Zone should take targeted measures:

- Hunshandake, Inner Mongolia, takes plant and engineering measures to combat desertification.
- The Maowusu desertification control area has restored natural vegetation to prevent dune activation and desert area expansion.

The soil erosion control area in the hilly and gully region of the Loess Plateau controls the development intensity. Take small watersheds as units, comprehensively control soil erosion, and build silt dams.

According to the requirements of the development direction of Inner Mongolia Forbidden Development Zone, there are mainly national nature reserves, world cultural and natural heritage, national key scenic spots, national forest park, and national geographic park.

The overall goal of the Inner Mongolia Autonomous Region during the 11th Five-Year Plan is to enhance the sustainable development of the social and economic environment. The natural population growth rate should be controlled within 7‰, and the total population should be controlled within 24.8 million. Cultivated land area has remained at 6.67 million hectares, ecological environment has been stabilized and curbed, key areas have been improved, vegetation coverage and grassland vegetation quality have been significantly improved, and forest coverage has reached more than 20%. In accordance with the principles of priority protection, active management, rational development, and intensive utilization, we will earnestly organize the implementation of key national ecological construction projects, intensify ecological protection and construction, and gradually improve the ecological self-repair capability. According to the principle of "who develops, who protects, who benefits, and who compensates," we will accelerate the establishment of an ecological compensation mechanism.

We will continue to implement key ecological construction projects such as Beijing-Tianjin sandstorm source control; returning farmland to forests, grazing, and grassland; natural forest protection; three-north shelterbelt; and soil and water conservation. We will increase scientific and technological support and improve the construction effect of the projects. We will further consolidate the achievements of ecological management in Horqin and Mu Us sandy lands and effectively improve the ecological conditions in Hulunbuir sandy land, Hunshandake sandy land, Wuzhumuqin sandy land, and Yinshan ecological pastoral areas. Protection and rational utilization of secondary forest resources such as Hulun buir grassland, Xilinguole grassland, Daxing'anling, and Alxa desert oasis. In Tengger, Badain Jilin, Ulanbuh, Baijin Windsor, and Kubuqi desert edge to establish windbreaks and sarin, to prevent the spread of the desert. Accelerate the implementation of the Yellow River in the upper and middle reaches of Inner Mongolia coarse sand area and Nenjiang River Basin black land. The construction of key forestry ecological projects has completed 4 million hectares, the area of soil erosion control has reached 2.1 million hectares, and animal husbandry has recovered 20 million hectares.

Vigorously develop forest industry, grass industry, and sand industry and other follow-up industries. In areas with both protection and timber conditions, we should vigorously develop fast-growing and high-yield forests and develop shrub industrial raw material forests and shrub feed forests. Actively develop forest products and forest product processing industry and forest characteristic breeding industry, and build a unique processing base for forest and grass products; we will guide and support leading enterprises in developing forest paper, wood-based panels, and shrubbery feed processing industries and build a forest product processing base in Daxing'anling forest region. To speed up the construction of artificial grassland, actively guide farmers and herdsmen to vigorously develop high-quality pasture. Sand industry focuses on developing desert tourism.

Strengthen the construction of forest and grassland fire prevention and forest pest protection and management system; strengthen the protection and management of wetlands, wildlife, geological relics, nature reserves, and geological parks; and effectively protect biodiversity. There are about 200 nature reserves of different levels. We will improve and implement the region's ecological function zoning, strengthen infrastructure construction in nature reserves of all levels and types, and protect and restore the ecological functions of key watersheds and regions.

4.1.3 Development Plans at the National and Autonomous Levels

The national economic and social development objectives during the 11th Five-Year Plan require that the trend of deterioration of the ecological environment be basically curbed. We will effectively protect the natural ecology, give priority to protection, and develop in an orderly way. Strengthen the ecological protection of natural resources such as water, land, forests, grasslands, and oceans. The focus is on promoting the protection of natural forests; returning farmland to forests, grazing, and grassland; controlling sandstorms in Beijing and Tianjin; controlling soil erosion, wetland protection, and desertification control of desert fossils; strengthening the ecological protection and management of nature reserves, important ecological function areas, and coastal zones; effectively protecting biodiversity; and promoting natural ecological restoration. According to the principle of "who develops, who protects, who benefits," we will accelerate the establishment of an ecological compensation mechanism.

The "11th Five-Year Plan" national development target is by 2010, the forest coverage rate will reach 20%, the forest stock volume will reach 132 billion cubic meters, the national ecological environment will basically stop, the ecological management in the western region will make breakthrough progress, and soil erosion in major river basins and desertification in major sandstorm areas will be greatly reduced. Medium-term target is by 2020, the forest coverage rate will reach over 23%. The ecological situation in the whole country has entered a virtuous circle, and the ecological situation in key western regions has improved significantly. Long-term goal is by 2050, the forest coverage rate will reach and stabilize above 26%, basically realizing beautiful mountains and rivers, complete forest ecosystem, and more developed forest industry system.

The key areas and projects supported by the central government in Inner Mongolia are as follows:

- Protection of natural forest resources.
- In the Yellow River basin, the conversion of cropland to forest and grassland continues in areas of soil erosion and sandstorm in the north.
- The eastern and western parts of Inner Mongolia are to control severely degraded grasslands
- Management of sandstorms in Beijing and Tianjin
- The fourth stage of the project is the construction of the "three north" shelterbelt system
- Wetland protection and restoration
- Soil and water conservation project

- Wildlife conservation and nature reserve construction to save endangered wildlife species
- Comprehensive management of rocky desertification areas

According to the "11th Five-Year Plan" of the state and the autonomous region and the requirements for the construction of new rural and pastoral areas, Inner Mongolia Autonomous Region has chosen 21 ecological construction projects, 73 water conservancy projects, and 22 farming and animal husbandry projects, with a total investment of 124.35 billion yuan.

Ecological construction projects focus on the development of forest industry, grass industry, sand industry and other follow-up industries. The implementation of Beijing and Tianjin sandstorm source control, returning farmland to forest, grazing and grassland, natural forest protection and other ecological construction projects, to consolidate the Horqin sand and Mu Us sand ecological management achievements. We will improve the ecological conditions in Hulunbuir sandy land, Hunshandake sandy land, Wuzhou Muqin sandy land, and Yinshan agropastoral ecotone and rationally develop and utilize forest resources in secondary forest areas such as Daxing'anling. In order to develop the grass industry, sand industry, and forestry disaster prevention and mitigation capabilities in conditional areas, a total of 21 projects were selected with a total investment of 54.01 billion yuan, accounting for 43.5% of the total investment in agriculture, animal husbandry, forestry, and water conservancy projects.

Water conservancy construction projects will focus on strengthening dikes and river courses of the main rivers and their tributaries, such as the Yellow River, Liaohe River, and Nenjiang River, and strive to build key water diversion and storage projects such as the Bera Estuary and Haibo Bay. We have completed a number of key water conservancy projects and dangerous reservoirs in Yue Le, Sanjiadian, and Sansheng Palace and have actively carried out the preliminary work of the "Diversion Project from Ji Liao" and "Diversion Project from Ji Xi in Harbin." In the construction of small- and medium-sized water conservancy projects in rural and pastoral areas, such as flood control, soil and water conservation, irrigation district construction, and water-saving renovation, 73 projects were selected with a total investment of 25.42 billion yuan, accounting for 20.4% of the total investment in agriculture, forestry, and water conservancy projects.

The construction of agricultural and animal husbandry infrastructure has strengthened the construction of bases. Agricultural and animal husbandry industrialization projects focus on supporting the construction of high-quality grain industry projects, large-scale commodity grain production bases, breeding projects, and agricultural and animal husbandry demonstration bases in areas where conditions permit. A total of 22 projects were selected, with a total investment of 44.92 billion yuan, accounting for 36.1% of the total investment in agriculture, forestry and water projects (Table 4.2).

Table 4.2 Indicators for phased control of land degradation control

	Near-term target (2006–2010)	Medium-term objectives (2011–2020)	Long-term goal (2021–2050)
Desertification control	The trend of land desertification and the stabilization and containment of the	The ecological environment of five sandy lands in Inner Mongolia is improved, and local	The desertified land that is suitable for treatment is basically fully controlled
	situation	optimization is achieved	Suitable forest land all greening
	Five-year increase forest area 40 million mu	Comprehensive containment of the trend of land desertification	Fundamental improvement of ecological conditions in sandy areas
	Forest coverage in sand areas increased to 10.5%	Forest coverage in sandy areas reached 14.6%	20% forest coverage in sandy areas
Conservation of water and soil	Increase of soil and water conservation comprehensive control area of 21666.7 km^2	Additional control area of 43,333 km^2	
	Construction of 2487 backbone projects in the Yellow River basin	Construction of 4974 key projects in the Yellow River basin	
	Soil and water conservation ecological restoration area 65,000 km^2	Soil and water conservation ecological restoration area 130,000 kg	
Forest	Artificial afforestation and afforestation of 30 million 700 thousand mu		
	Aerial planting 13 million mu		
	Closed mountain (sand) forest 24.82 million mu		
Biodiversity	Establishment of 5 highland biodiversity nature reserves		
	Strive for 5 international wetland directories, 3 wildlife conservation and nature reserves, 7 national nature reserves, 9 new antelopes, brown bear, chelated and crane nature reserves, 3 new breeding bases for red deer, and other provenances		
	More than 13% of nature reserves		

4.1.4 Consistency Between Inner Mongolia Autonomous Region and National Land Degradation Control Plan

The National Action Plan to Combat Desertification has been drafted and concentrated in 265 key counties in the western region. The three stages will be 2001–2010, 2011–2030, and 2031–2050. An important objective of the first phase of the NAP is to manage 22 million hectares of degraded land by 2010.

The Plan of Action for Biodiversity Conservation is formulated to implement the Convention on the Protection of Biological Diversity. The National Biodiversity Research Report compiled an index of endangered plants through comprehensive evaluation of biodiversity and put forward policy recommendations to strengthen national biodiversity protection and sustainable use of biological resources.

In order to implement the International Convention on Biological Diversity and incorporate the protection and sustainable use of biological diversity into its national strategy and action plan, China began to prepare the China Biodiversity Protection Action Plan in 1992. The "Action Plan" was officially launched in June 1994. The plan of action has determined the ecological location and list of priority species for biodiversity conservation in China and has set targets in 7 areas, including 26 priority action plans. In addition, according to the urgency and feasibility of protection, 18 priority projects need to be implemented immediately.

Forestry development planning. Forestry development in the new century will include the protection of natural forest resources, the conversion of farmland to forests, the construction of key protective forest systems in the three norths and the Yangtze River Basin, the management of sandstorms in Beijing and Tianjin, the protection of wildlife, and the construction of nature reserves. Based on the framework of six major forestry projects such as the construction of fast-growing and high-yield forest bases in key areas, a national forest ecological network system combining "points, lines, and surfaces" is constructed, i.e., national urban greening areas, forest parks, surrounding nature reserves, and typical ecological areas are "points"; major rivers, mountains, coastlines, and railway lines are "lines." In order to achieve a balanced and reasonable distribution of forest resources, eight regions, namely, the northeast state-owned forest region in Inner Mongolia, the arid and semiarid regions in northwest and western northeast China, the North China Plain and the Central Plains Plain region, the southern collective forest region, the southeast coastal tropical forest region, the southwest mountain canyon region, and the Qinghai-Tibet Plateau alpine region, are taken as "areas".

National wetland conservation action plan. In order to promote the implementation of the provisions of the Ramsar Convention, the Convention has been ratified. This includes designating important wetlands as national protected areas.

The Inner Mongolia Autonomous Region has formulated relevant plans, selected planning objectives, and formulated protection and governance plans in accordance with the national planning framework such as the Eleventh Five-Year Plan for National Economic and Social Development in the Inner Mongolia Autonomous Region. The project scale and implementation steps are proposed.

4.1.5   *Western Development Strategy and Land Degradation Control*

In the early stage of the western development, Inner Mongolia began to implement key ecological construction projects such as natural forest protection, returning farmland to forests, controlling sandstorms in Beijing and Tianjin, and returning grazing to grassland. The scope of construction covers more than 90 counties and cities in the region.

Strategic priorities: soil erosion and the construction of sand and salt control areas in the middle and upper reaches of the Yellow River, the construction of sand source control areas in Beijing and Tianjin, the construction of Hulun buir and Xilinguole grassland protection control areas, the construction of conservation zone in Daxing'anling natural forest, the construction of Alashan natural fence management area, and the construction of environmental protection areas.

Key projects: grassland ecological construction and protection project, natural forest resource protection project, returning farmland to forest and grass project, comprehensive ecological environment control project, sand prevention and control project, "three north" shelterbelt project, soil and water conservation project, green passage project construction, environmental protection project, and nature reserve and ecological demonstration area construction project.

By the end of 2005, the total construction scale had exceeded 270 million mu, 35.04 million mu of farmland had been converted to forests, and 150 million mu of grass had been planted. Pushed by key ecological construction projects, the first is to achieve a major breakthrough in controlling an area larger than that of desertification. According to the third desertification monitoring in 2004, compared with 1999, the desertification land in the whole region decreased by 24 million mu and 4.8 million mu each year, and the desertification land decreased by 7.3 million mu, with an average decrease of 1.46 million mu. The ecological deterioration in Horqin sandy land has been under key control with a forest coverage rate of 20%. The second is the initial restoration of grassland ecology. The area of degraded grassland in Inner Mongolia is 820 million mu, and the total area of grassland protection and construction in 2005 is 750 million mu. Third is the trend of soil erosion has weakened. In recent years, the accumulated soil erosion area in Inner Mongolia has reached 126 million mu, accounting for 56% of the soil erosion area. The amount of sediment from the Ulanbuh Desert to the Yellow River dropped from 230 million tons to 190 million tons per year. Forest coverage in Inner Mongolia has increased from 13.81% in 1997 to 17.5% at present (Table 4.3).

Table 4.3 Conversion of farmland to forests in Inner Mongolia

	Reuse farmland	Barren mountain wasteland	Close hillsides to facilitate afforestation	Total
2000	50	127		177
2001	80	90		170
2002	367	600		967
2003	457	537		994
2004	160	500		660
2005	216	180	140	536
Total	1330	2034	140	3504

Unit: 10,000 mu

4.2   Investment in Land Degradation Prevention and Poverty Alleviation Since 1985

4.2.1   Input from Central Ministries of Inner Mongolia Autonomous Region

Since the western development, the central government has increased its investment in Inner Mongolia year by year, which has promoted the economic and social development of Inner Mongolia. From 2000 to 2005, the central government allocated more than 176 billion yuan to Inner Mongolia for financial transfer payments, special subsidies, and construction bonds and 23.23 billion yuan to ecological projects, accounting for 13.2% of the total central investment. In the western region, the state has carried out key ecological construction projects, such as returning farmland to forests, returning to grassland, soil and water conservation, natural forest protection, and Beijing-Tianjin sand management. Overall, all key projects are progressing smoothly and have achieved remarkable results.

In addition, according to the unified arrangement of the State Council, 19 central government agencies have been designated to support 26 key counties in Inner Mongolia in carrying out national poverty alleviation work. Beijing supports Inner Mongolia counterpart. From 1997 to 2004, Beijing municipal finance provided 186 million yuan of support funds and 192 million yuan of donated materials to poor areas in Inner Mongolia. Beijing has implemented 728 poverty alleviation projects in Inner Mongolia, helped build 705 km of roads, built 89 schools, and jointly organized 100 "sister schools," benefiting 670 teachers in poor areas of Inner Mongolia. Beijing and Inner Mongolia are making joint efforts to help the poor, improve production and living conditions in some poor areas of Inner Mongolia, promote economic development in poor areas, and solve the problem of food and clothing for nearly a million poor people. At the same time, from 1996 to 2004, the cooperation between enterprises in Beijing and Inner Mongolia also developed rapidly. Enterprises of both sides implemented more than 1700 cooperation projects with a capital of 35.685 billion yuan (Table 4.4).

Table 4.4 Central input of ecological projects (2000–2005)

	Total	2000	2001	2002	2003	2004	2005
Ecological key flag county	19,340	15,640	3700	–	–	–	–
Natural forest resource protection project	514,698	63,552	133,409	100,502	83,635	67,317	66,283
Conversion of cropland to forest and grass	743,740	8000	33,200	150,930	144,040	199,178	208,392
Beijing-Tianjin sand source control project	527,719	10,600	93,310	158,080	8280	166,480	90,969
Return grazing land to grassland	202,746	0	0	0	43,227	47,674	111,845
Soil and water conservation[a]	30,601	600	4600	3400	2924	7400	11,677
Three North Forest Protection Project	14,840	–	5840	1000	1000	1000	6000
Conservation and construction of natural grassland	25,130	–	6250	14,800	4080	–	–
Remote poverty alleviation and relocation	102,000	–	15,000	30,000	50,000	70,000	–
Forbidden grazing and drylot feeding	94,493	–	–	9256	21,631	31,803	31,803
Forest ecological benefit compensation fund	47,460	–	–	–	–	23,710	23,750
Total	2,322,767	98,392	295,309	467,968	358,817	551,562	550,719
Implementation area							
Ecological key counties	Scope of implementation: 10 municipalities, with the exception of Hulunbuir and Xingan, and 29 flag districts						
Natural conservation project	Involving Forest Industry Group, Lingnan eight Bureau, and the Yellow River basin 29 flag counties						
Shelterbelt engineering	12 league cities and 66 county cities						
Erosion and torrent control works	12 league cities and 68 county cities						
Conversion of cropland to forestry project	12 league cities and 96 county cities						
Beijing-Tianjin sand source control project	4 league cities and 31 county cities						

(continued)

Table 4.4   (continued)

	Total	2000	2001	2002	2003	2004	2005
Natural grassland protection and construction project	6 league cities and 27 county cities						
Natural forest resource protection project	9 league cities and 27 county cities						
Returning grazing to grass project	6 league cities and 27 county cities						

Source: Financial Statistics and Analysis of Inner Mongolia
Unit: 10,000 yuan
[a]Soil erosion control and silt dam works

4.2.2   Input from Governments in Inner Mongolia Autonomous Region

In 2005, Inner Mongolia's total income was 33.505 billion yuan, while local fiscal expenditure was 73.471 billion yuan, with a difference of -39.966 billion yuan. Inner Mongolia's total local fiscal budget revenue is 27.745 billion yuan (belonging to local government revenue), while Inner Mongolia's total local fiscal budget expenditure is 68.165 billion yuan (excluding special funds from the central government). The local financial self-sufficiency rate is 40.70, and the balance of revenue and expenditure mainly depends on central transfer payments. Due to the low degree of marketization in the western region, the government has undertaken too many social functions and public expenditures. It is not easy to reduce the scale of government expenditure, and the pressure to increase expenditure is still great. Therefore, the per capita budget expenditure of western provinces and cities is generally higher than that of developed provinces and cities.

Since 1985, the Inner Mongolia government has invested about 1 billion yuan in poverty alleviation funds through various departments in the autonomous region, of which about 200 million yuan has been spent on forest and grass industries related to land degradation, water conservancy development, and poverty alleviation in other places. Of the total investment of 200 million yuan in the past 3 years, about 30 million yuan has been used to prevent land degradation, 33.2 million yuan has been allocated to local poverty alleviation funds each year, and 15.56 billion yuan has been allocated to local finance. In 2005, the central government and the autonomous region invested a total of 5.51 billion yuan in the project of returning farmland to forest and grassland, banning grazing and grassland, and protecting natural forests in Inner Mongolia, of which 2.7 billion yuan was used for grain and cash subsidies for farmers and herdsmen and compensation funds for public welfare forests. However, according to the budget arrangement of Inner Mongolia at the beginning of the fiscal year, the budget funds for agricultural resources and environmental protection are 2014 million yuan, 28.73 million yuan for returning farmland to forests, 5 million yuan for forest ecological benefits and afforestation, 2.7 million yuan for preventing and controlling desertification, and 880,000 yuan for natural forest protection projects. Therefore, most of the funds for preventing land degradation and poverty alleviation come from the central government. Local governments are constrained by their financial capacity. The annual investment for preventing land degradation and poverty alleviation is very limited (Table 4.5).

4.2.3 Input from Local Governments in Inner Mongolia Autonomous Region

At present, Inner Mongolia's local finance is divided into four levels: autonomous region, union city, flag county, and hematoxylin township, "level 1 government, level 1 administrative power, level 1 financial power, and level 1 budget." The financial difficulties of counties and townships are a common problem in the whole country, and the financial problems of counties and townships in Inner Mongolia are no exception. The main reasons are the adjustment of national and local fiscal and tax systems, the reform of taxes and fees in rural and pastoral areas, and the influence of tax reduction and exemption policies in the western development policy. Local taxes are relatively reduced, and the proportion of county financial resources is generally low. The extremely high financial support for population in the counties

Table 4.5 Comparison of general budget receipts and expenditures of local finance in Inner Mongolia and Northwest China (2005)

	Finance income of local administration (100 million)	Rank	Per capita budget with general income (yuan/person)	Rank	Finance income of local administration (100 million)	Rank	Per capita budget with general income (yuan/person)	Rank
Inner Mongolia	277.45	20	1163.79	10	681.65	16	2859.27	6
Shanxi	274.99	21	742.21	19	641.10	18	1730.37	18
Xinjiang	180.32	26	918.59	12	517.02	24	2633.84	10
Gansu	123.38	27	471.08	29	428.04	27	1634.35	20
Ningxia	47.71	29	811.35	16	169.68	29	2699.70	8
Qinghai	33.76	30	626.26	21	158.68	30	4147.98	5

Notice: Guangdong (180.6 billion yuan), Shanghai (141.7 billion yuan), and Jiangsu (132.2 billion yuan) ranked the top three in general budget revenue of local finance
Guangdong (CNY 228.8 billion), Jiangsu (CNY 165.2 billion), and Shanghai (CNY 164.6 billion) ranked the top three of the general budget expenditure of local finance
Shanghai (8137 yuan), Beijing (6157 yuan), and Tianjin (3239 yuan) ranked the top three in per capita income
Shanghai (9450 yuan), Beijing (7039 yuan), and Tibet (6768 yuan) ranked the top three per capita expenditure

is also the main reason for the disjointed financial operation of the county government and the increasingly serious financial problems. Although governments at all levels have repeatedly stressed that they should try their best to transform government functions and improve the environment for economic growth, the government's long-term fiscal deficit has eroded the potential for future economic development. However, investment by local governments in combating land degradation and poverty is limited financially (Table 4.6).

4.2.4   Input from Donor Organizations in Inner Mongolia Autonomous Region

From 1980 to 2005, the Inner Mongolia Autonomous Region signed 113 foreign-funded loan projects with a total investment of 47.5 billion yuan. Of this amount, the loan agreement amount was 2.694 billion US dollars, and the actual foreign capital used was 1.919 billion US dollars, accounting for 71.2% of the total contractual foreign capital. From the perspective of project implementation, most projects can complete the construction progress according to the project plan and reach the expected goals. International agencies have different approaches to land degradation management and poverty reduction in Inner Mongolia. Activities cover various types of projects in most sectors, including microcredit, small-scale infrastructure development, community development, environmental protection, technical assistance, capacity building, and integrated rural development. The World Bank and the Asian Development Bank are relatively large in scale and have also provided technical assistance to help Inner Mongolia improve its ability to reduce land degradation and poverty.

From 2001 to 2005, Inner Mongolia signed a total of 40 foreign loan projects with a loan agreement amount of US$ 802.1054 million, adding an average of 8 foreign loan projects every year. Environmental protection, ecology, poverty alleviation, and other projects have become the main areas of foreign loans. It became the characteristic of this period.

World Bank loan project: Inner Mongolia borrowed 38 projects from the World Bank. The loan contract amount of the World Bank was US$ 1.101 billion, and the actual use amount was US$ 828 million, accounting for 75.22% of the total. Among them, seven education projects, three water conservancy projects, one poverty alleviation project, and five health projects were completed on schedule. According to the assessment of the World Bank, the overall situation of the project is good, and the rural water supply and sanitation project loaned by the World Bank in Inner Mongolia is well received. The soil and water conservation project and tuberculosis control project in the Loess Plateau of Inner Mongolia won the World Bank President Award.

Asian Development Bank loan project: Inner Mongolia used the Asian Development Bank to borrow four projects, with contractual loans of US$ 140 million, and

Table 4.6 Changes in fiscal revenues and expenditures at different levels in Inner Mongolia

	Center		Municipality		League City		Banner		Township	
	Raise	Refund	Income	Outcome	Income	Outcome	Income	Outcome	Income	Outcome
1990			3.94	12.20	10.90	17.46	13.48	25.10	4.65	6.14
1991			4.87	14.39	13.13	18.97	16.82	32.23	4.58	6.35
1992			3.99	16.48	13.92	19.96	15.88	28.42	5.29	7.22
1993	–	28.32	4.64	18.81	22.83	26.38	22.02	33.61	6.63	9.48
1994	31.92	28.84	5.54	23.87	10.77	23.69	12.80	23.69	7.18	10.63
1995	32.64	29.06	6.22	24.97	12.60	25.54	15.68	39.24	9.20	12.43
1996	35.98	29.98	6.18	27.75	16.44	31.83	19.13	50.68	15.50	16.12
1997	38.01	30.50	7.00	31.30	18.68	35.86	22.02	57.46	18.38	18.30
1998	41.46	31.27	9.64	40.46	21.05	43.39	27.98	66.29	18.98	20.16
1999	42.87	31.59	10.22	54.97	23.35	49.27	32.39	73.34	20.61	22.22
2000	44.91	32.04	12.56	77.88	25.95	57.40	35.81	87.45	20.71	24.53
2001	50.38	–	14.98	90.75	27.64	76.51	40.72	119.44	16.09	32.56
2002	–	–	18.29	97.33	28.29	105.15	50.04	152.04	16.24	39.06
2003	–	–	19.47	97.86	28.83	115.16	69.94	192.22	20.48	42.02
2004	–	–	26.61	112.73	40.09	137.80	103.14	259.59	26.91	54.00
2005	–	–	37.31	139.75	45.71	151.69	155.00	318.21	39.44	72.22
2000–2005 rate of rise(%)			31.64							

Source: Inner Mongolia financial statistics report
100 million yuan

actually used US$ 114 million, accounting for 79.9% of the total. The emergency rehabilitation and reconstruction project in northeastern Inner Mongolia and the Songhua River flood control management project in Inner Mongolia are still under construction.

Yen loan project: In Inner Mongolia, 14 loans were made in Japanese yen, while Japan's Bank for International Cooperation made a loan of 85.293 billion yen, with 527 million US dollars actually used, accounting for 60.2% of the contract amount. The afforestation project and the air pollution and environmental control projects in Hohhot and Baotou have been completed. The sandstorm control project in Inner Mongolia, the talent education project in Inner Mongolia, the water environment control project in Hohhot, the atmospheric environment control project in Baotou, and the infrastructure project in Tongliao are all under construction.

International IFAD Loan Project: Inner Mongolia uses the International Fund for Agricultural Development (IFAD) to carry out loan projects, with a loan contract amount of 12.44 million US dollars and an actual use amount of 11.64 million US dollars, accounting for 93.56% of the contract amount. The grassland animal husbandry development project in northern China is located in the northern part of Chifeng City, including Alukorqin flag, Balinyou flag, Keshkeid flag, and Wuniute flag. The construction period of the project began in 1981 and ended at the end of December 1988. The total investment was 69.18003 million yuan, accounting for 96.71% of the investment plan.

Foreign government loan projects: There are 61 projects in Inner Mongolia funded by foreign governments, with loan contract amount of 360 million US dollars and actual use amount of 321 million US dollars, accounting for 85.8% of the contract amount. After the implementation of most projects, the economic and social benefits are good, and the equipment is in good condition. However, due to various reasons, some projects did not meet the expected income after being put into production, and some project units stopped production or even went bankrupt, increasing the local financial burden.

International commercial loan project: There are five projects in Inner Mongolia that utilize foreign capital for commercial loans, with a contract value of US$ 190 million. The actual use of US$ 117 million, accounting for 62% of the contract amount, has all been put into production (Table 4.7).

4.2.5 Investment from Private Enterprises and Non-governmental Organizations in Inner Mongolia Autonomous Region

International non-governmental organizations include the Global Environment Facility, Ford Foundation, World Wide Fund for Nature, Oxfam Hong Kong, German Bank for Reconstruction, Eide Foundation, World Vision, Save the Children

Table 4.7 Loan status of international financial organizations and foreign governments (up to 2005)

Project	Loan agreement money	Terms of loan				Loan balance
		Time	Interest rate	Commitment rate	Poundage	
Total	189431.3					104251.9
World Bank	77761.0					38441.3
Northern Irrigation Project	6799.5	20(5)^a	4%	0.50%		1623.0
The First Stage of Loess Plateau	3724.9	20(8)	3%	0.75%		1404.0
The Second Stage of Loess Plateau	2483.0	17(5)	Floating interest rate	0.75%		2483.0
Western Poverty Alleviation Project	6100.0	17(5)	0.75%	0.50%		5900.0
Poverty Education in Ethnic Areas	1757.9	20(5)	1.20%	0.50%		1673.9
Technical Assistance for Poverty Reduction	25.0	Grant				0.0
Western Distance Network Education	25.0	10(5)	0.75%	0.50%		0.0
Others						
…						
Asian Development Bank	13905.0					11299.98
Waterborne Reconstruction Project	11000.0	30(7)	Floating interest rate	0.75%		11000.0
Songhua River Flood Control Project	2805.0	25(5)	Floating interest rate	0.75%		300.0
New Energy Alternatives in Poor Areas	50	Grant				
Energy Development Research	50	Grant				
United Nations Fund for Agricultural Development	1607.0					1393.8
Northern Grassland Project	1607.0	50(10)	1%	0.75%		1393.8
Bilateral Government Loan Project	96158.4					53116.9
Afforestation Project	3300.0	40(10)	0.75%		0.25	2130.4
Sand Control Project	12661.0	40(10)	0.75%		0.15	1739.1
…	…	…				…

Unit: 10,000 dollars
Source: Inner Mongolia Financial Statistics and Analysis
^{a}The figures in brackets refer to the grace period for loans

UK, Plan International, International Crane Foundation, and American Progressive Organization. International non-governmental organizations' poverty alleviation activities are small in scale and in various forms. Some focus on the combination of environmental protection and poverty reduction, while others focus on community development and local capacity building. In addition, the Hong Kong Kadoorie Foundation's assistance in the second phase is 3 million yuan. The China Foundation for Poverty Alleviation and the Inner Mongolia Foundation for Poverty Alleviation have jointly invested 40 million yuan per year for remote consultation since 2005. The medical technology and equipment project and the China Education Foundation have invested 680,000 yuan in the construction of two hope primary schools in Junggar and Yijinhuoluo Banners. In 2005, the State Foreign Investment Project Management Center led CCED to invest 8 million yuan in Wuniute Banner of Chifeng City to carry out community development projects and support poor farmers to carry out comprehensive poverty alleviation.

Domestic non-governmental organizations for poverty alleviation include the China Association of Poverty Alleviation Foundations, the China Charity Federation, the All-China Women's Federation for Poverty Alleviation, the All-China Disabled Persons' Poverty Alleviation Federation, Project Hope, Glorious Cause, happiness project, and Soong Ching Ling Foundation.

The China Poverty Alleviation Fund will mainly carry out the following activities: microcredit, capacity building, promotion of practical technologies, emergency rescue, maternal and child health care, and primary education. The activities of China Charity Federation mainly include supporting income-generating activities, living assistance, medical assistance, teaching assistance, and vocational training. The activities of the All-China Women's Federation for Poverty Alleviation include scientific and technological training, microcredit, twinning, labor export, girls' education, small infrastructure, and women's health care. Activities carried out by the All-China Disabled Persons' Federation include practical technical training, microcredit, reconstruction of dilapidated buildings, and construction of service cooperatives. The main activities of Hope Project include providing grants to out-of-school children, building Hope Primary School, teacher training, and teaching equipment. Glory career activities include project investment, donation for running schools, and other public welfare undertakings. The main activities of the happiness project are microcredit. Soong Ching Ling Foundation's poverty alleviation projects mainly include girls' aid, primary and secondary school construction, mobile libraries for children, grants for female normal school students, teacher training, and teacher incentive funds.

Compared with the government's poverty alleviation, non-governmental poverty alleviation is relatively small in scale and generally covers only some poor areas. However, non-governmental poverty alleviation activities are generally more specialized and concentrated in the private sector, which has advantages in professional fields. For example, on June 5, 2004, the Alxa SEE Ecological Association (SEE), funded by nearly 100 well-known Chinese entrepreneurs, aims to improve and restore the ecological environment in the Alxa region of Inner Mongolia and to slow down and contain the occurrence of sandstorms and gradually expand the scope of control to other desertification areas in China while promoting more eco-logical responsibilities of entrepreneurs. The Alassane Ecological Association, which operates on the basis of non-governmental organizations, received a donation of 2.4 million pounds. For example, the Hope Project supports primary education in poor areas, while happiness project provides microcredit to women. Non-governmental poverty alleviation activities are relatively more creative and efficient.

Since 1985, the private sector and non-governmental organizations have invested about 180 million yuan in the autonomous region to prevent land degradation and poverty (Table 4.8).

4.3 Lessons Learned from Land Degradation Control Projects

4.3.1 Scope, Activities, and Outputs of Past and Existing Integrated Ecosystem Management and Land Degradation Control Projects in the Autonomous Region

Table 4.8 Projects related to preventing land degradation implemented in Inner Mongolia

Main activities	Main output
1. The Beijing Tianjin sandstorm source control project (Wulanchabu, Xilun, Chifeng, and Baotou, 4 cities, 31 banners)	
Fencing and improvement	Desertification degradation control area is 11.309 million mu
Forage base and artificial grassland	
Aerial seeding and artificial supplementary seeding	The total scale of grassland construction is 74 million mu
Prohibition and cessation of grazing	The area of small watershed management is 4 million 524 thousand mu
	Water source engineering and water-saving irrigation (16,656 sites)
2. Conversion of farmland to forests (12 cities, 96 banners)	
Steep slope cultivated land	13.3 million mu
Serious desertified cultivated land for conversion	Barren mountain wasteland is 21.74 million mu
Carrying out food instead of relief	
Recommendation of planting trees, restoration of vegetation	
3. Natural forest resource protection project (key state-owned forest area, Lingnan Baicui, middle and upper Yellow River engineering area)	
Create a public welfare forest, a commercial forest	17 million 368 thousand mu of public welfare forest construction
Diversion and placement of laid-off workers	Forest management area up to 175.526 million mu

(continued)

Table 4.8 (continued)

Main activities	Main output
Reduction of forest cutting	Full placement of surplus staff and workers
Protection of natural forest resources	Reduce timber output by 530,000 m³
4. "Three north" shelterbelt construction project (12 cities, 66 banners, and urban areas)	
Create shelterbelts for farmland, pastures, river basins, roads, railways, and towns	Afforestation (artificial, flying, sealing) with 10 million mu
5. Fast-growing and high-yield forest construction project (Yellow River, Liaohe River, and Nenjiang River basins)	
Development of fast-growing and high-yielding timber forest	Construction of 1 million mu of fast-growing and high-yielding forest
6. Wetland, wildlife conservation, and nature reserve construction project (district)	
Establishment of wetland conservation coordination mechanism and management system	36 wetland nature reserves have been expanded and constructed, with a total area of 3 million hectares
Preliminary formation of wetland protection network system	At the end of 2005, the number of nature reserves reached 188, and the area of nature reserves reached 14.244 million ha
7. Key public welfare forest protection projects (6 cities, 41 banners)	
Complete the definition of forest classification and regionalization in Inner Mongolia	Public welfare forest area with 47 million mu
Construction of the first batch of key public welfare forests launched in an all-round way	
8. "Kapok corridor in northern Xinjiang" (on both sides of the corridor of arrong banner and arzhuo banner) forestry ecological management project	
All barren hills, wasteland, and hillsides within sight have been afforested	Tree planting with 13.8 million Mou
9. Natural grassland protection project (9 cities, 27 banners)	
10. The project of returning pasture to grass (6 cities, 27 banners)	
Rotational grazing system in typical grasslands and Meadow grasslands in the Central and Eastern part of China	100 million mu of forbidden grazing area, 140 million mu of off-grazing area, 10 million mu of rotational grazing area
Forbidding grazing in degraded sandy areas or on the edge of desert sandy land	250 million mu of returning pasture and grass project
In the moderately degraded sandy grassland, grazing is the main thing, and the combination of prohibition and rest is the main one	
11. Construction of monitoring supervision system for grassland disaster: prevention and mitigation	
Strengthening grassland Supervision in an All-round way	Autonomous region grassland ecological remote sensing dynamic monitoring and information analysis and processing center
Establishing and perfecting the fire prevention system of grassland	

(continued)

Table 4.8 (continued)

Main activities	Main output
Establishing and perfecting the Reserve of Fire Prevention material in grassland	Construction of autonomous region level 1 grassland fire prevention command center and 8 municipal grassland fire prevention command centers
	Establish and improve 7 fire prevention material reserves on the alliance's municipal grasslands
	The establishment of an autonomous district level rodent damage forecast station, 10 municipal level rodent damage forecast stations, 25 county level rodent damage forecast stations
12. Grassland and pastoral area water conservancy project	
Development of Water-saving Irrigation forage Base	Water Saving Irrigation forage Base with 2 million mu
	20,000 wells of drinking water for people and animals in pastoral area
13. Soil and water conservation engineering (12 league cities, 68 banners and municipalities)	
Key management projects of national soil and water conservation	Key Control Project of silt Dam in Loess Plateau
Recommendation of planting trees, Construction of basic farmland and Construction of Channel Control Project	World Bank loan Project for soil and Water Conservation in the Loess Plateau
	Soil and Water loss Control Project in Sand and coarse Sand area
	Pilot Project of Integrated Prevention and Control of soil and Water loss in Black soil area of Northeast China
	Soil and Water Conservation Ecological Restoration pilot Project
	Construction Project of soil and Water Conservation Ecological Environment in Heihe River
14. Environmental protection project (upper Nenjiang River, upper reaches of Xiliao River, Western foot of Daxingan Mountains, Northern foot of Yinshan Mountain, Ejina Banner, Hulunbuir City, Aohan Banner of Chifeng, Daqingshan, Dalai Lake, Horqin, Helan Mountain, Daqinggou)	
Air pollution, water pollution control and nature reserve, construction of ecological demonstration zone	Five special ecological functional areas
	Construction of 15 Ecological demonstration Banner, County and 72 Ecological demonstration areas
	Construction of 73 National and Autonomous Regional Nature Reserves
	Four environmental monitoring and information system construction projects

4.3.2 Successes and Lessons Learned from Past and Existing Projects

The successful experience of Inner Mongolia in past land degradation control and ecological construction projects is mainly reflected in the following aspects:

Mobilizing the whole society to participate in ecological protection and construction: Local governments attach great importance to the construction of ecological environment, firmly grasp the opportunity of the central government to implement the western development strategy, take the ecological construction as the starting point of the autonomous region, and put forward the strategic decision of "ecological zone." Party committees and governments at all levels should improve their organizational structure, be responsible to the people, guide farmers and herdsmen to join in ecological construction, and promote sustainable, rapid, healthy, and coordinated development of ecological construction in the whole region. The ecological environment in Inner Mongolia has been greatly improved.

Strengthen scientific and technological support and increase scientific and technological contribution rate: In many years of ecological construction, Inner Mongolia has accumulated many valuable experiences. ABT rooting powder, water retention agent, water gun, plastic film mulching, and other technologies can effectively improve the survival rate of afforestation. At the same time, the science and technology department participated in different projects and adopted the form of science and technology supporting demonstration areas to carry out engineering demonstration and promote the application of new technologies, increasing the proportion by three. Scientific and technological achievements should be organically combined with ecological engineering to improve the contribution rate of science and technology.

Strengthening project supervision to ensure project quality: All major ecological construction projects in Inner Mongolia are supervised by successful supervision companies, eliminating some hidden dangers and ensuring project quality.

Perfect the system and control risks: Inner Mongolia has carried out a series of systematic construction of key ecological construction projects, such as bidding system, supervision system, acceptance system, and capital audit system, to ensure the construction quality and investment benefits.

Lessons learned: International experience clearly shows that the multifactor method is conducive to ensuring the long-term success of land degradation control. In the field of land degradation, the government has learned the following lessons, which will help prevent land degradation in the future.

Institutional coordination and planning harmonization: There is an urgent need to strengthen inter-agency coordination and cooperation in policies, legal systems, projects and budgets for land degradation prevention and control at the central to local levels. At present, there are more than a dozen offices in the autonomous region related to the prevention and control of land degradation. The existing coordination mechanism clearly shows that it is difficult for an institution to coordinate activities at other levels at the same level. An important lesson is that under the current management mechanism and cultural habits, only through clear functions and mutual cooperation can effective interdepartmental coordination be realized.

Consistent legal and regulatory framework: Although it is emphasized that the responsibility for planning and implementation should be delegated to the lowest possible level, there are still a lot of duplication and overlap in environmental legislation to be solved. In the short term, the soil and water conservation law needs to be revised as soon as possible. Another priority is the implementation of the new antidesertification law. In the medium term, a separate and integrated approach to the conservation and management of natural resources is needed. Another priority is the implementation of the new antidesertification law. In the medium term, a separate and integrated approach to the conservation and management of natural resources is needed.

Decentralization of implementation responsibilities: Recently, the autonomous region government has taken a series of steps to devolve more administrative responsibilities to local governments, such as cities, thus creating opportunities that need to be actively seized. The coordination and rationalization of natural resources management requires the reform of existing departmental obligations and responsibilities and the strengthening of the capacity of local governments to assume more responsibility for implementation. Broad participation of interest groups in decision-making plays an important role in promoting local governments to manage natural resources, reducing institutional barriers, and developing successful projects.

A stable policy to fundamentally solve the problem: We should strengthen the understanding, identification, solution, and monitoring of the root causes of land degradation, rather than merely solving superficial problems. The root cause of the problem is the root cause of unreasonable methods of land use and management, including the socioeconomic environment of rural land users (poor farmers, farmers, and herdsmen), as well as their social, economic, cultural, and policy environment (such as population pressure, poverty, inappropriate promotion, or farmers not fully mastering). Modern agricultural technology, unreasonable development policies, insufficient service quality, and unreasonable legal environment.

Sustainable method: Land degradation, especially land use planning, has been developed and applied to standardized and sometimes inappropriate land management methods and technical programs, regardless of geographic, ecological, social, and economic conditions. Engineering measures (ecological construction) are still the most important way to solve ecological problems. However, in the long run, a participatory and integrated approach at the ecosystem level is more effective and sustainable than an investment-dependent engineering approach. Actions to address land degradation should be consistent with existing important national plans and projects in order to obtain strong national and autonomous support. Unilateral actions, tree planting, or sand dunes are usually ineffective.

Strengthening monitoring and evaluation: Monitoring and evaluation of previous projects are not enough, and their governance effects cannot be judged as good or bad. Previous agencies involved in monitoring and analyzing land degradation trends have adopted different definitions of land degradation or have focused on only one aspect of the problem (e.g., soil erosion). An effective coordination program and standardized procedures are needed to collect information on land degradation trends so that decisions can be made based on accurate, timely, and widely accepted data. Through the application of integrated ecosystem management methods, monitoring and evaluation can effectively reflect the effect of project management.

4.3.3   Costs and Benefits of Past and Existing Projects

At present, the state has implemented key ecological construction projects in Inner Mongolia, including returning farmland to forest and grassland, soil and water conservation, natural forest protection, and sandstorm source control in Beijing and Tianjin. In general, key projects are progressing smoothly with remarkable results. However, the economic and ecological benefits are different. Due to the different years of project implementation and adopted measures, horizontal comparison is very difficult. According to the unified index system, we sorted out and briefly analyzed the following items. That is to calculate the direct benefits (increase the output value of grass, increase the output value of trees and increase the output value of crops) of the added value of the main products after the project is invested, and use the indirect benefits to reflect the ecological benefits (carbon sink value, oxygen generation value, etc.).)。 Soil and water conservation value, etc.), the life span is 10 years. Regardless of changes in interest rate and discount rate, the final result is reflected by Ibi (comprehensive income index) and sorted by this index. The calculation method and procedure are as follows:

$$IBI = \Sigma WB$$

In the above formula, the weight and score of each index are obtained.

Regarding the weight of economic benefits and ecological benefits, we use economic benefits and ecological benefits to calculate the average value of economic benefits and ecological benefits, respectively. The economic benefits of the project are analyzed in the form of benefit-cost ratio, which is directly expressed by numerical value with a weight of 0.6. Ecological benefits are evaluated by experts with a weight of 0.4:

$$IBI = \text{Economic Benefit value} / \text{Cost} \times 0.6 + \text{Ecologcal Benefit value} \times 0.4$$

According to the relevant requirements and procedures for economic evaluation of construction projects in China, quantitative and dynamic analysis methods shall be adopted for economic benefit analysis as far as possible. The specific steps are as follows:

(a) Arrange items and collect project data.
(b) The project is further subdivided into several subprojects.

(c) Determine the project's input (n) and discount rate (6%).
(d) Detailed calculation of costs, fixed costs and variable costs, mainly input costs, operating costs, interest, and so on.
(e) Calculate benefit. It is divided into direct income and indirect income.
(f) B/C.

Ecological efficiency, using expert evaluation method, invite experts (12–18 people) who are familiar with the prevention and control of land degradation to rate the ecological efficiency of the project, taking into account the consistency with the budget/cost analysis value, the smaller value can be used to score out of 5 points; For the scores evaluated by experts, these two points are removed and the average value is taken with a weight of 0.4. Judging from the investment in the project, the central government has taken on too much responsibility for transfer payments. The central government has further increased the transfer payment after the country completely reduced the tax on agriculture and animal husbandry. This has also led to local government rent-seeking activities, operating projects, and increasing subsidies becoming the daily work functions of local government officials, which to some extent reduces the motivation of local government to control expenditures.

Judging from the implementation of foreign aid projects, the scale is from small to large, and the area involved is also expanding. Most foreign loan projects can complete the construction progress according to the requirements of the project plan during the construction period and reach the expected goals. In particular, poverty alleviation projects have played a vital role in helping farmers and herdsmen get rid of poverty and become rich. However, some preliminary work on the project is not enough. In order to compete for new projects and reduce consideration of objective conditions, some investment consulting and design units are making preparations for the initial stage of the project to meet the urgent requirements of local governments and units for the introduction of projects, funds, and their own economic interests. The feasibility study and demonstration of the project cannot be carried out objectively, scientifically, and impartially, which leads to long running time and poor implementation effect in the early stage of the project. Judging from the management system of foreign aid projects, the administrative department lacks supervision and follow-up evaluation on the overall planning and implementation of foreign aid projects. Foreign loans are "borrowed, used, and repaid" out of touch, and the debt service mechanism is not perfect. Causes the project unit to use the fund positively, does not repay the loan voluntarily the phenomenon; foreign loan statistics are not perfect, and local foreign loan projects lack comprehensive statistics.

As most of the land degradation projects and ecological projects are public welfare projects, debt service pressure is very high. At present, the peak period of repayment for external loan projects in Inner Mongolia Autonomous Region has arrived. However, because the direct economic benefits of public welfare projects themselves are not high, and many basic projects are ultimately directed to farmers, it is more difficult for governments at all levels to recover funds. In addition, project financing capacity is poor. International Monetary Fund loan projects require 50% of the support funds, while domestic bank loans, as one of the main channels to support investment, are not easy to fully implement. Due to limited financial resources, the matching funds promised by the local authorities cannot be put in place in time.

However, in order to win the project, the local government promised to support the loan first, but it was difficult to reach the project in full and on time, which affected the progress and efficiency of the project.

4.3.3.1 Case 1: China Western Poverty Alleviation World Bank Loan Project

Case 1: Loan Project of the World Bank for Poverty Alleviation in Western China

In July 1997, Inner Mongolia Autonomous Region was designated as the World Bank Poverty Alleviation Project Area in Western China. According to the 17-year soft loan term, the total amount of loans was 27.586 million special drawing rights (soft loans) and 23.3 million US dollars (hard loans). Including a 5-year grace period and a 20-year hard loan period. The project covers the northern foot of Yinshan Mountain in Inner Mongolia, the hilly and gully region on the central and southern edge of the Loess Plateau, and the edge of Horqin sandy land, covering 6 alliance cities, 15 counties, 137 hematoxylin villages and towns, 910 villages and towns; the new project covers 4 alliance cities, 12 counties, 47 hematoxylin villages, and 253 Gacha villages. The project consists of 6 subprojects and 62 sub-projects. The specific subprojects and sub-projects are as follows: (1) subproject 37 on land and farmer development; (2) sub-item 15 on irrigation and land development; (3) 4 subprojects for rural infrastructure construction; (4) Subproject 2 of rural enterprises; (5) health subitem 1; and (6) subitem 3 of institution building. The main objectives of the project are as follows:

1. Significantly reduce the absolute poverty level in 21 counties with poor families and to increase the per capita disposable income of the poor in the project area.
2. Improve the labor skills and quality of beneficiaries (families) through effective implementation of projects. Change some traditional and inefficient family management methods.
3. Promote the integration of local natural resources through the implementation of the project, increase agricultural production and income, and promote the economic growth of the project area.
4. Encourage most farmers to participate in the project by playing a decisive role in the design and implementation of the project.
5. Through the construction of irrigation and land development, infrastructure and sanitation projects, the living environment in the project area can be effectively improved, and the deterioration of the ecological environment in poor rural areas can be curbed.
6. Strengthen poverty alleviation organizations, train project management teams, improve the management level of poverty alleviation projects,

(continued)

monitor the poverty level, and accumulate and enrich the management experience of other poverty alleviation projects in the future.

The total investment of World Bank loan projects is 1024 million yuan, of which 506.7 million yuan is World Bank loan and 519.3 million yuan is domestic loan.

Project results: (1) The absolute poverty population in the project area decreased from 784,200 to 249,000. (2) The per capita net income of the project households was 1862 yuan, and the per capita grain consumption was 760 kg, exceeding the target for the project period. Compared with 1999, the per capita livestock ownership increased from 3.15 sheep to 4.83 sheep. (3) The production and living environment in the project area has been greatly improved, mainly in solving the problems of power supply and drinking water in rural areas. (4) All project farmers have received training related to the project, improved their labor skills, and mastered 1–2 applicable production technologies.

The project lasted for 7 years. After overall acceptance evaluation, the project planning and design were more reasonable, and the established project objectives were completed while ensuring the project quality. Moreover, the implementation of the project conforms to the design standards, has achieved obvious benefits, and has popularization applicability in technical application; the demonstration effect of the project is strong, which has promoted the development of local industrialization, scientific project management, and far-reaching social benefits.

Statement of completion of investments by subproject

Project name	Object	New funds	Medium-term adjustment target	Accomplished	Proportion of mid-key target %
Land and farmer development	42423.85	13222.09	67800.68	76405.11	113
Irrigation and land development	16098.75	3857.22	20063.99	16233.53	81
Rural infrastructure	5201.92		1842.72	2047.64	111
Hygienism	2000.00		381.34	371.34	97
Rural enterprises	2225.20	161.64	3067.60	2898.49	94
Institution building	2403.23	332.05	2716.94	3278.87	121
Flat cost	70352.95	332.05	2716.94	101234.98	3726
Total cost	83000.00	17573.00	95873.26	101234.98	106

Unit: 10,000 yuan

Case 2: UNICEF SPPA Project

Wushen Banner is located in Inner Mongolia Autonomous Region and is supported by the United Nations Children's Fund SPPA project. After successfully completing the project cycle objectives from 1996 to 2000, it also experienced a transition period from 2001 to 2003. During the phased implementation of the SPPA project from July 1996 to October 2003, Ushenqi received 1.892 million yuan, 400,000 yuan for training, and 36 sets of special equipment from 4 small loan funds of the United Nations Children's Fund (UNICEF). Local governments invested 3.586 million yuan in matching funds, of which the flag government invested 2.346 million yuan and the township government of Sumu invested 1.239 million yuan. In addition, 28 pieces of various local special equipment were used, with special training funds and 674,000 yuan for public services. From 1996 to 2003, the loan balance was 11.25 million yuan, and the microcredit funds operated well, which promoted the social and economic development of Wushen Banner.

The project funds are relatively concentrated; the group has fixed deposits on a regular basis, established a repayment guarantee system, and basically perfected the corresponding financial management system. The SPPA Project in Wushen Banner has achieved good financial results.

An analytical table on the organization and financial operation of the transition period of the SPPA project from 2001 to 2003 in Ushenqi

	Unit	2000	2001	2002	2003
Project organization					
1. Number of project villages and towns	Individual	12	12	10	7
2. Number of projects	Individual	39	39	19	18
3.Number of women members	Individual	3443	1689	1208	1089
4.Number of women in effective loans	Individual	3258	1361	973	943
5.Cumulative number of women loans	Individual	3443	4804	5777	6720
Microfinance fund investment					
1. UNICEF input	Ten thousand yuan	189.2	189.2	189.2	189.2
Local government matching fund					
1.Matching funds of flag government finance	Ten thousand yuan	90.0	0	0	0
2.Town government finance matching funds	Ten thousand yuan	123.97	0	0	0
3. Training funds	Ten thousand yuan	25.0	3.5	5.0	3.7
4. Official expenses	Ten thousand yuan	20.0	10.1	8.0	6.0

(continued)

	Unit	2000	2001	2002	2003
5. Other earmarked funds	Ten thousand yuan	9.6	8.0	9.4	10.0
Financial operation					
1. Annual loan amount	Ten thousand yuan	50.0	189.6	199.9	210.0
2. End loan balance	Ten thousand yuan	190.0	194.0	200.9	210.5
3.Accrued interest income on projects	Ten thousand yuan	33.0	45.87	63.97	84.17
4. Overdue loan amount	Ten thousand yuan	3.8	8.0	0.9	0
5. Amount of loan recovered in the current period	Ten thousand yuan	186.2	186.0	199.0	200.9
6. Repayment rate	%	98.0	96.0	99.5	100.0

Source: According to the final evaluation report of SPPA project of the people's government of Wuhenqi banner in Inner Mongolia Autonomous Region

4.3.4   Methods and Technologies for Land Degradation Control

In view of the land degradation and ecological environment problems in Inner Mongolia, most of the methods and technologies used in the past were single projects, but there are still deficiencies in comprehensive aspects. According to the overall strategy of ecological construction in our district, in terms of land use and prevention measures, we should adhere to the methods of unified planning, reasonable layout, key construction, and comprehensive management and adopt mature technologies and successful methods.

1. The eastern comprehensive management area is mainly based on the comprehensive methods and technologies of ecological environment management. Mature technologies, experiences and methods of natural forest protection, three-north shelterbelts, sand prevention and control, grassland construction, and soil erosion control have been adopted.

   The Daxing'anling primeval forest region mainly relies on the "natural protection project" to strengthen the construction and protection of key projects. The former secondary forest area has established a family-based forestry management system; developed green food, especially in farming and wood product processing industries; carried out diversified operations to increase the income of forest people; and established a new type of Linyang forest protection and ecological benign production relationship.

Water conservation forests, mountain farming forests, and soil and water conservation forests have been built in the Nenjiang and Erguna river basins to protect biodiversity, improve agricultural conservation tillage, degrade the ecosystem, and protect the water resources in the upper reaches of northeast old industrial bases. Develop fisheries and eco-tourism, take the path of ecological civilization, and build a strong ecological economic zone.

2. Keerqin grassland, Hulunbeier grassland and key sandy land have been fenced. Water conservancy projects in pastoral areas and the construction of artificial forage grass lands have been carried out simultaneously. Accelerate the development of animal husbandry production and animal husbandry processing industry, strengthen artificial recommendation (aerial seeding) and grassland improvement, strengthen soil erosion control measures and desertification control in black soil areas, and plant sandy shrub belts and forest belts. Harness and protect black soil resources, promote the sustainable use of water and soil resources, and ensure ecological safety.

   In the central part of Inner Mongolia, the comprehensive control technology of sand prevention and control is mainly applied in Beijing and Tianjin area, supplemented by the introduction of new technologies. This area is a national key ecological project construction area, which is implemented in 31 counties of 4 allied cities. The main focus of this area is to prevent and control desertification, and ecological methods such as forest cultivation, aerial seeding, and fencing have been adopted. Water-saving irrigation, variety optimization, transformation of medium- and low-yield fields, and development of featured agriculture using facility farming and animal husbandry technologies. Make full use of wind energy and heat energy, wind power generation, development of biogas, development of green energy, improve the production and lifestyle of local farmers and herdsmen.

3. Comprehensive measures and related technologies for soil erosion and desertification land management in the middle and upper reaches of the yellow river.

   This area is a soil erosion area caused by wind erosion, water erosion, gravity erosion, and sand salinization, involving 56 counties such as Mu Us sandy land, Kubuqi Desert, and Ulanbuh Desert. The main measures and technologies for comprehensive management of the ecological environment include:

   First, strengthening the natural forest resource protection project, protecting the existing secondary forest and plantation resources, establishing and improving ecological protection and related policies and measures, and encouraging and supporting the whole society to participate in and invest in ecological construction.

   Second, in combination with the project of returning farmland to forest and grassland, farmland with steep slopes and farmland with serious wind erosion should be returned to forest.

   Third, in combination with the project of returning grazing land to grassland, we will increase the construction of grassland fences and infrastructure in pastoral areas.

Fourth is to form a comprehensive management system from slope to ditch to reduce the Yellow River sediment by combining the key control projects of warping dams on the Loess Plateau, water and soil conservation projects, and Yellow River control projects.

Fifth, *Hippophae rhamnoides* and *Caragana korshinskii* soil and water conservation forests should be built in areas with the most serious arsenic pollution.

Sixth, in combination with the "three north" shelterbelt project and the sand prevention and control project, we will strengthen the management of the Mu Us Desert, Kusu Desert, and Ulanbuh Desert and develop tourism and sand industry.

Seventh, select suitable areas and use precipitation conditions to carry out aerial seeding and enclosure in time. Combined with fencing and returning farmland to forests, areas with water conservancy conditions will be selected to build stable and high-yield basic farmland.

4. Management methods and techniques of sandstorm source area in Alashan. Precipitation in arid desert areas of Alxa League is very small (the lowest annual precipitation is only 50 mm), which is the main source of sandstorms. The main measures and technologies for comprehensive ecological management are as follows: (1) to transfer people and animals in areas with serious ecological deterioration in a planned way in combination with ecological construction projects. (2) In combination with the ecological environment construction of water and soil conservation in Heihe River, increase the water distribution in the lower reaches of Heihe River to ensure the sustainable storage of water resources and the improvement of water environment in the East and West Juyan Sea. (3) The existing *Populus euphratica* forest, *Haloxylon ammodendron* forest, *Tamarix chinensis* forest, and Helan mountain desert oasis in Juyan Sea have been protected and restored. (4) Establish a management and protection system for nature sanctuaries, replace the status of herdsmen in sanctuaries, enforce the prohibition of grazing and keeping in captivity, and enable herdsmen to receive pension and become keepers of sanctuaries. Properly resettle ecological migrants, train migrants, transfer to towns and mining areas, and transfer to places with better ecological conditions. To actively develop Cistanche, organic green food industry, adjust the industrial structure, based on the artificial cultivation technology of Haloxylon ammodendron, to ensure the survival of the artificial cultivation of Haloxylon ammodendron. The ecological benefits of *Haloxylon ammodendron* and the economic benefits of *Cistanche deserticola* are realized through the comprehensive application of sand fixation technology, *Haloxylon ammodendron* artificial planting technology, and *Cistanche deserticola* cultivation technology.

5. Methods and techniques for grassland degradation control in pastoral areas of Inner Mongolia. We should protect natural vegetation, strengthen fences, improve grass seeds, and implement artificial grass planting, aerial grass planting, and rotational grazing. Implement seasonal animal husbandry, strengthen animal husbandry infrastructure construction, and improve the ability of disaster prevention. Promote biological control technology for grassland rodent damage,

grazing chickens to control grasshopper type C virus rodent damage, eagle rack deratization, etc. Maintain the ecological balance of grasslands, control weasel and mouse infestation, and control wild animal pests.

Techniques for preventing and controlling land degradation through agricultural ecosystems. The purpose of preventing land degradation in farmland ecosystem is to reduce soil erosion and land salinization and to maintain and improve land productivity. Inner Mongolia's main technologies in this area include:

Conservation tillage of farmland: Conservation tillage of farmland is one of the effective measures to prevent land erosion in arid areas. In Inner Mongolia, the area of low-yield farmland accounts for 40% of the total arable land and is dry farming. The main measures of conservation tillage are to make full use of comprehensive agricultural mechanization technology, promote straw returning to the field, improve soil fertility, and prevent soil erosion.

Increasing organic fertilizer and planting green manure: Animal husbandry in Inner Mongolia is developing rapidly. Animal husbandry production in agricultural areas accounts for more than 50% of animal husbandry in this area. In addition, the price of agricultural resources has increased in recent years, and the proportion of organic fertilizer used in agricultural production has also increased. This improved the soil in Inner Mongolia and reduced the loss of soil nutrients. Reducing soil compaction and improving the quality of cultivated land play a positive role. Planting and applying green manure can improve soil salinization. Tillage, intercropping, green manure, rotation, and stubble retention of green manure and crop varieties are conducive to slowing down the salinization of land.

Farmland infrastructure: Farmland infrastructure effectively curbs land degradation. Through scientific planning of farmland infrastructure areas, forests, roads, canals, and basic water conservancy facilities, not only has the unit output of farmland been increased but also the land degradation has been effectively curbed.

Rural energy construction: Rural energy construction directly affects the control of land degradation and the improvement of ecological environment, reduces energy consumption, and contributes to the protection and growth of forest and grass vegetation. Rural energy construction is a reconstruction project of straw returning to the field and energy-saving biogas digester (also called tricalcium pool). The rural energy construction project has been popularized and implemented in Inner Mongolia for more than 20 years and has been welcomed and supported by farmers. Using ecological principles and systematic scientific methods and using modern technology and traditional methods, through reasonable investment and clever combination of time and space, the ecological system can maintain benign material and energy circulation and realize the coordinated development of human and nature. It is divided into five categories: soil conversion technology, vegetation restoration and reconstruction technology, land degradation control technology, small watershed comprehensive control technology, and land reclamation technology.

Chapter 5
Land Degradation and Its Prevention in Inner Mongolia

5.1 General Situation of Land Resources

Inner Mongolia Autonomous Region is located at the northern border of the motherland and extends from northeast to southwest, taking a long and narrow shape. Now it has jurisdiction over 3 leagues, 9 cities and 101 flags, counties, cities and districts. The east-west line is about 2500 kilometers long and the north-south line width is about 1700 nm. The total land area of the whole region is 1.183 million square kilometers, accounting for 12% of the national land area, ranking third in the country. The per capita land area of the whole region is about 0.05 km².

Inner Mongolia Autonomous Region straddles east and west, with significant climate differences. Based on the mountains of Daxing'anling, Yinshan, and Helan Mountains, there are obvious regional differences in land resources and their utilization in the whole region. In the north of the mountain, high plains (land) are mainly composed of desert and sand, mainly used for animal husbandry. Grassland accounts for 85.08% of the land area and is an important animal husbandry production base in our country. In the south of the mountain area, there are mainly hills, terraces, and plains. There are also some deserts. Land use is intertwined with agriculture and animal husbandry. The proportion of cultivated land, pasture, and woodland in the land area is 12.6%, 52.77%, and 20.11%, respectively, forming an important agro-pastoral ecotone. Among them, the eastern, central, and western plains are dominated by irrigated agriculture, while the mountainous areas are dominated by forestry. Judging from the quality and suitability of land resources, the east is better than the west and the south is better than the north.

Climate types in Inner Mongolia Autonomous Region are complex and varied. Most areas have temperate continental climate. Only the northern part of Daxing'anling is a mountainous continental cold climate. In winter, the whole area is controlled by Mongolian high pressure. In summer, the influence of southeast monsoon gradually weakens from southeast to northwest. Langshan and western Helan Mountains are controlled by continental air masses. At the same time, the

© Science Press & Springer Nature Singapore Pte Ltd. 2020
Z. Meng et al., *Public Private Partnership for Desertification Control in Inner Mongolia*, https://doi.org/10.1007/978-981-13-7499-9_5

barrier function of the southwest Qinghai-Tibet Plateau prevents warm and humid air from entering the Indian Ocean, making the climate drier. Hydrothermal climate factors show a northeast-southwest trend. From the northern part of Hulun Buir to the western plateau of Alashan, the temperature exceeds 10 °C, 1400 °C rises to 3600 °C, and the annual precipitation change rate is an important factor affecting land resources. The annual precipitation in Inner Mongolia Autonomous Region is generally between 50 and 450 mm, decreasing from east to west. The rainfall in the westernmost part is only 25 mm.

The main landform types in Inner Mongolia are plateau, mountain, hill, plain, desert, and terrace. They are distributed in a belt shape throughout the region, with a trend from northeast to southwest. Geomorphic features are unique. Geomorphic types have a regular transition from Daxing'anling and Yinshan to both sides. It has obvious zonality. As far as the whole region is concerned, the drought is aggravated from east to west, the water flow in the east is prominent, and the river network is dense. Most of the western regions show seasonal floods, which eventually turn into drought and erosion. Wind power is very active in the arid and semiarid regions of western China and plays a leading role in erosion, transportation, accumulation, and destruction of land forms. The desert (sand) area is very large and tends to expand.

Various climatic conditions and geomorphic features have provided conditions for the formation of various disasters and also become inducing factors of land degradation, such as land desertification, soil erosion, salinization, etc.

5.2 The Situation of Land Degradation

From an ecological point of view, land degradation refers to the deterioration of plant growth conditions and the decline of land productivity. From the perspective of system theory, land degradation is the result of interaction between human factors and natural factors. In essence, the basic connotation and change process of land degradation are reflected in soil degradation, including physical degradation, chemical degradation, and biological degradation. In recent years, the word "soil degradation" has been widely used to replace land degradation in the world. However, land degradation and soil degradation are different after all. Replacing land degradation with land degradation is not comprehensive. After all, land is a natural complex composed of rock, landform, climate, hydrology, and biology with a certain thickness, and its structure and function far exceed the scope of soil. Land degradation is a very complicated and comprehensive dynamic process, which contains a very strong concept of time. The so-called degradation and non-degradation should be understood by comparing the quality and quantity of land in different periods, such as deserts, Gobi, snow plains, and some stone desert areas. The quality and quantity are the same for a long time. Therefore, it is impossible to include degraded land. To be exact, in the comparison period, the quality (quantity) of land in the later period is obviously lower than that in the earlier period. Land environment is land degradation.

According to the actual situation in Inner Mongolia Autonomous Region, land degradation should be defined by the United Nations Convention to Combat Desertification (UNCCD). In other words, land degradation refers to the biological or economic productivity of rainwater, irrigation or grassland, pasture, forest, and woodland in arid, semiarid, and dry subhumid areas due to the use of land or due to the combination of battalions or several battalions. Complexity is reduced or even lost.

The Inner Mongolia Autonomous Region occupies about 1/8 of the land area. The vast land area and complex and diverse natural environment provide greater possibility and diversity for the formation of various natural disasters, especially in the mid-latitude region, which is characterized by multiple disasters and multiple difficulties. Land degradation is serious in Inner Mongolia. Desertification is spreading at a rate of more than 4% in seven areas, including three areas in Inner Mongolia. The desertification area of Inner Mongolia Autonomous Region has reached 62.24 million square kilometers, ranking second only to Xinjiang. The direct economic loss caused by desertification in the country is about 54 billion yuan and the indirect loss is 170 billion yuan, with Inner Mongolia Autonomous Region accounting for one-third each.

At present, there is no unified plan for the classification of land degradation types at home and abroad, but most researchers mainly classify the causes and consequences of land degradation. Inner Mongolia Autonomous Region has a vast territory, and the types of land degradation are relatively complex. The classification mainly adopts two-level system to describe the types of land degradation in Inner Mongolia Autonomous Region. The first stage is the differentiation of dynamic factors leading to land degradation. The second level is divided by vegetation type, i.e., ecosystem. According to the driving factors of land degradation, the main types of land degradation in Inner Mongolia Autonomous Region are:

Wind erosion refers to the phenomenon that surface soil substances are transported away from the surface due to the action of wind and the abrasion of particles in the airflow to the surface. According to the bulletin on desertification and desertification issued by the forestry department of the autonomous region in July 2005, the climate in the Inner Mongolia Autonomous Region is dry and windy. Desertification is mainly caused by wind erosion. The land with wind erosion desertification is the most widely distributed, covering 563,000 square kilometers, accounting for 47.62% of the total area of the autonomous region and 90.52% of the total area of desertified land (Fig. 5.1).

Water erosion is defined by soil migration and deposition caused by atmospheric precipitation, especially rainfall. The Inner Mongolia Autonomous Region Forestry Department issued the "Inner Mongolia Autonomous Region Desertification and Desertification Bulletin" in July 2005. Soil erosion and desertification land in Inner Mongolia Autonomous Region is mainly distributed in hilly and mountainous areas in the central and eastern part of the autonomous region, with an area of 27,500 km², accounting for 4.43% of the total area of desertification land.

Salinization refers to the accumulation process of harmful soluble salts in soil. Groundwater and surface water are harmful to plants. According to the Inner

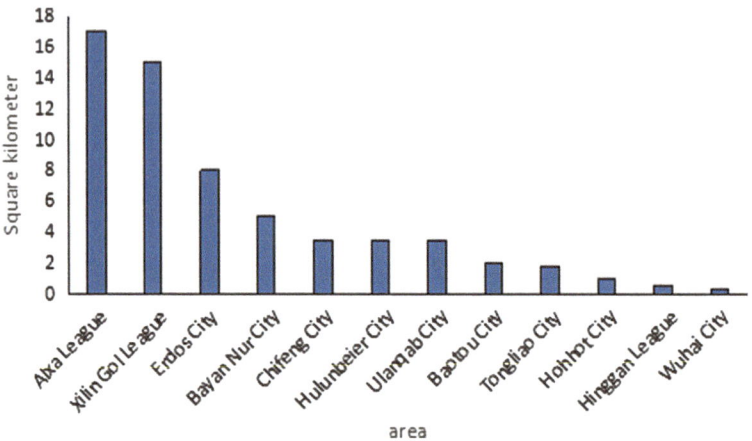

Fig. 5.1 Current situation of desertification land in the Inner Mongolia Autonomous Region

Mongolia Autonomous Region Desertification Status Bulletin issued by the Inner Mongolia Autonomous Region Forestry Department in July 2005, salinized land in Inner Mongolia Autonomous Region is less distributed, mainly concentrated in Hetao Plain, Tumerchuan Plain, and hilly lowlands of several sandy lands, with an area of 31,300 km², accounting for 5.05% of the total area of desertification land.

Freeze-thaw refers to the damage caused by soil mechanics when the temperature changes at or below 0 °C. The freeze-thaw erosion area in this area is 47,699 km², accounting for 3.98% of the total area. According to the remote sensing survey of erosion in 2000, the area of soil freeze-thaw erosion in this area is slightly higher than 47712.71 km², accounting for 4.03% of the total area of the autonomous region and accounting for the three types of erosion areas of wind, water, and frost (5.84% of the area with slight erosion above). From the results of the two statistics, the freeze-thaw erosion area in Inner Mongolia Autonomous Region has increased, but the increase is very small, which is mainly caused by climate. Freeze-thaw erosion is mainly distributed in the Daxing'anling region in the northeast of the autonomous region, i.e., the permafrost distribution area in Daxing'anling, which is approximately consistent with the distribution of cold and warm coniferous forests in the mountainous region of the autonomous region.

Other factors causing land degradation include the exploitation of mineral resources; the erosion of projects such as urban and traffic road construction; the use of pesticides, fertilizers, and plastic films by agricultural production; and the soil pollution caused by wastes and garbage from industrial enterprises.

According to the ecosystem classification of Inner Mongolia Autonomous Region, the types of ecosystem degradation are as follows: forest ecosystem degradation, grassland ecosystem degradation, desert ecosystem degradation, wetland ecosystem degradation, and farmland ecosystem degradation.

5.3   Land Degradation Prevention and Control Work

5.3.1   Central Government

Since the western development, the central government's investment in Inner Mongolia has increased year by year, promoting the economic and social development of Inner Mongolia. In recent years, Inner Mongolia's central government transfer payments and special subsidies for ecological projects account for 13.2% of the total central investment. Western countries have implemented measures such as returning farmland to forests and grasslands, soil and water conservation, natural forest protection and sandstorms in Beijing and Tianjin. Ecological construction of key projects. On the whole, the key points of the project are progressing smoothly and the results are obvious.

In addition, according to the unified plan of the State Council, 19 central state organs have designated 26 key national poverty alleviation counties in Inner Mongolia. To help the counterpart poverty alleviation and Inner Mongolia poverty alleviation form a joint force, improve the production and living conditions in poor areas of Inner Mongolia, promote the economic development of poor areas, and solve the problem of food and clothing for nearly one million poor people. While carrying out counterpart cooperation, the cooperation between enterprises in Beijing and Inner Mongolia has also developed rapidly.

5.3.2   Autonomous Government

Due to the low degree of marketization in the western region, the government has taken on too many social functions and public expenditure, the scale of government expenditure is not easy to reduce, and the pressure of expenditure expansion is still great. Therefore, the per capita expenditure in the western provinces is higher than that in the developed provinces and cities.

Over the past 30 years, about 18% of the total investment of the Inner Mongolia government has been used to prevent and control land degradation. Most of the funds for land degradation and poverty alleviation have been invested in the central government. The local financial capacity is restricted, and the annual investment in land degradation and poverty alleviation is very limited.

5.3.3   Local Government Departments

At present, Inner Mongolia Autonomous Region's finance is divided into four "upper level government, first level power, first level property right, and first level budget" counties and villages. Financial difficulties are common problems, and

Inner Mongolia's county and township finance is no exception. The main reason is that with the adjustment of national and local tax systems, the reform of rural taxes and fees, and the impact of tax reduction policies in the western development policy, local taxes are relatively reduced, and the proportion of fiscal revenue from self-owned sources in counties is generally low. On the other hand, due to the impact of welfare expenditures such as public works construction, wage adjustment, and social security, the scale of expenditures is expanding rapidly, and the finance is drying up day by day, resulting in a higher level of government transfer payments and increasing dependence on payments. The main reasons are the extremely high dependence on finance, the disjointed financial operation of the county government, and the increasingly serious financial problems. Although governments at all levels have repeatedly stressed the need to change government functions and improve the environment for economic growth, the long-term existence of government budget deficits has eroded the potential for future economic development. Under such circumstances, the local government has a very limited financial position to prevent land degradation and poverty.

5.3.4 Donor Agency

The Inner Mongolia Autonomous Region and foreign contracted foreign loan projects to control land degradation, from the perspective of project implementation, most of the project planning requires the completion of construction progress and the achievement of the expected goals. International institutions on land degradation in Inner Mongolia differ in governance and poverty reduction activities, covering most sectors of various types of projects, including microcredit, small-scale infrastructure construction, community development, environmental protection, technical assistance, capacity building, and comprehensive rural development. The World Bank and the Asian Development Bank are relatively large in scale and help Inner Mongolia improve its ability and efficiency in land degradation and poverty alleviation through technical assistance.

World Bank Loan Project: According to the World Bank's assessment, the overall situation of the project is good. Among them, the World Bank Loan Project for Rural Water Supply and Environmental Sanitation in Inner Mongolia has won favorable comments, and the Phase II Soil and Water Conservation Project and Tuberculosis Control Project in Loess Plateau of Inner Mongolia have won the World Bank President Award.

Yen Loan Projects: Inner Mongolia, Hohhot, and Baotou have completed projects on air pollution and environmental management using yen loans.

IFAD Loan Project: The IFAD Loan Project in Inner Mongolia is a grassland animal husbandry development project in northern China. The project area is distributed in 15 hematoxylin (fields) in the northern part of Chifeng City, including Alukolqin Banner, Balinyou Banner, Keshiketeng Banner, Ongniute Banner, and 4 flags.

Foreign Government Loan Projects: Most of the projects have good economic and social benefits, can repay debts, and have good equipment. However, some projects have not achieved the expected benefits due to various reasons. Some project units have stopped production or even gone bankrupt, increasing the burden on local finance.

The Inner Mongolia International Commercial Loan Project, which uses foreign commercial loans, has been put into full operation.

5.3.5 *Private Enterprises and Non-governmental Organizations*

International NGOs mainly include the Global Environment Facility, Ford Foundation, World Wide Fund for Nature and Oxfam Hong Kong, Friendly Foundation, World Vision, Save the Children UK, Plan International, International Crane Foundation, American progressive organization, etc. International non-governmental organizations' poverty alleviation activities are small in scale and take various forms. Some focus on the combination of environmental protection and poverty alleviation, while others focus on community development and local capacity building. In addition, a second aid package of three million yuan will be provided to the Kadoorie Foundation in Hong Kong. The China Foundation for Poverty Alleviation and the Inner Mongolia Foundation for Poverty Alleviation will invest 40 million yuan annually from 2005 to jointly provide remote consultation, medical technology, and equipment projects. The China Education Foundation will invest 68 million yuan to rebuild the two Hope Primary Schools in Zhungeer and Ejina Banner. In 2005, led by the State Foreign Project Management Center, CCed Project in Ongniute Banner of Chifeng City invested eight million yuan to carry out community development projects to support poor farmers to carry out comprehensive poverty alleviation.

Domestic non-governmental relief organizations mainly include China Foundation for Poverty Alleviation, China Charity Federation, All-China Women's Federation Women's Poverty Alleviation Action, CDPF Poverty Alleviation Activities, Hope Project, Glorious Cause, Happiness Project, Soong Ching Ling Foundation, etc.

Activities to be carried out by China's Poverty Alleviation Fund mainly include microcredit, capacity building, promotion of practical technologies, emergency rescue, maternal and child health care, and primary education. The activities of China Charity Association mainly include supporting income-generating activities, living assistance, medical assistance, student assistants, and vocational training. Activities carried out by the National Women's Federation for Action against Poverty: science and technology training, microcredit, twinning, labor export, girls, small infrastructure, and women's health. Activities of the national federation of poverty alleviation are practical technical training, microcredit, reconstruction of dangerous houses, and service construction. The main activities that want to carry out this project include grants for dropouts, construction of Hope Primary School, teacher training, and provision of teaching equipment. Guang Cai's activities mainly include project investment, running schools, and other public welfare undertakings. The main activities of the project are microcredit. The projects of Soong Ching Ling Foundation for poverty alleviation mainly include female students, construction of primary and secondary schools, mobile libraries for children, teacher training for female students, teacher training, teacher incentive fund, etc.

Compared with the government's poverty alleviation, the characteristics of non-governmental poverty alleviation are relatively small, generally covering only some poor areas. However, folk activities are generally more specific, focusing on the advantages of private organizations in professional fields. As of June 5, 2004, Alxa SEE Ecology Association has been established with the support of nearly 100 well-known Chinese entrepreneurs. The aim is to improve and restore the ecological environment in Alxa region of Inner Mongolia, alleviate and contain the occurrence of sandstorms, and gradually expand the scope of control to other parts of the desert region and, at the same time, promote entrepreneurs to assume more ecological responsibilities. Alxa SEE Ecology Association received a donation of 2.4 million pounds in accordance with the operation of non-governmental organizations. Another example of Hope Project primary education is mainly to support poor areas, and Happiness Project mainly provides microcredit for women.

Chapter 6
Land Degradation Action Plan in Inner Mongolia

6.1 Priority Area

6.1.1 Priority Area for Land Degradation Control Operations

Inner Mongolia has a special geographical location, with little rain and drought. More than two-thirds of the region's ecological environment is fragile and sensitive. The climate is getting drier and drier. Land and surface vegetation are seriously degraded. Sandstorm disasters are intensifying. Various disasters are frequent. Sandstorm is rampant and the area of land desertification is continuously expanding. The intensification of the ecological crisis has not only caused great losses and serious difficulties to the economic development and people's lives in Inner Mongolia Autonomous Region but also posed a serious ecological threat to the "three northern" regions of our country, especially Beijing, Tianjin, and their surrounding areas. The consequences have seriously affected the development of social, economic, political, and military undertakings in northwest, north, northeast, and even the whole country. Therefore, harnessing and improving the ecological environment; establishing Inner Mongolia's ecological defense system; achieving sustainable economic, social, and ecological development; and reconstructing beautiful mountains and rivers have become major strategic issues that the country urgently needs to solve. At present, there are various problems in different ecosystems in Inner Mongolia. Among them, grassland degradation, soil erosion, and desertification are the main components of Inner Mongolia's ecological environment. The formation of ecological environment problems in Inner Mongolia has a profound historical origin and is closely related to the sensitive changes of many modern natural factors and the participation of human activities. Climate change and unreasonable utilization of human resources development are important factors leading to deterioration of ecological environment. According to the current situation of degradation, the possibility of restoration, and the necessity of taking preventive measures to prevent land degradation, the ecosystem of Inner Mongolia

© Science Press & Springer Nature Singapore Pte Ltd. 2020
Z. Meng et al., *Public Private Partnership for Desertification Control in Inner Mongolia*, https://doi.org/10.1007/978-981-13-7499-9_6

Autonomous Region is listed as a priority. The first thing to be protected is farmland ecosystem, followed by swamp ecosystem, forest ecosystem, meadow ecosystem, grassland ecosystem, and desert ecosystem.

6.1.1.1 Comprehensive Control Area in Eastern Inner Mongolia

This includes Daxing'anling forest region, Hulun Buir grassland and sandy land, Nenjiang River Basin, and Erguna River Basin, involving parts of Hulun Buir City, Xingan League, Xilinguole League, and Tongliao City. There are many ecosystem types in this area, including forest ecosystem, grassland ecosystem, farmland ecosystem, and wetland ecosystem. The Great Xing'an Mountains are 1300 km long from the north to the south, forming the Nenjiang and Liaohe rivers. Nenjiang River Basin used to be an area with abundant water resources and dense forests. For a long time, due to natural, economic, social, and historical reasons, forest resources have been severely damaged, forest margins have shrunk, and forests have been destroyed. Especially in the areas where agriculture and forestry intersect, excessive reclamation and deforestation and overgrazing lead to forest degradation, dwarfism, and residual degradation. Forests reduce water resources and maintain water and soil functions, resulting in serious water and soil loss from south to north in Nenjiang River Basin. Drought and flood disasters intensified and species became extinct, causing great disasters to animal husbandry production and the lives of farmers and herdsmen in Nenjiang River Basin. This area is also an important water source for the economic revitalization and social development of the old industrial base in northeast China. At the same time, this area is also an important production base for commodity grain and animal husbandry in our country. The sensitive areas of soil erosion in this area are mainly distributed in the hills with low vegetation coverage and the three grassland areas of Hulun Buir, Xilinguole, and Horqin. Desertification-sensitive areas are mainly distributed in the arid and semiarid areas of the grassland in the west of Hulun Buir, Xilinguole grassland, and Horqin Sandy Land, while the land salinization-sensitive areas are relatively small, mainly distributed in the arid and semiarid areas along the river and inland plateau basins. Focus on black soil, natural forests, nature reserves, sand control, grassland construction and prevention of soil erosion, construction of soil and water conservation forest, and farmland protection forest.

6.1.1.2 Wind-Sand Source Area in Central Inner Mongolia

It is located in the middle of Inner Mongolia, including 31 counties of 4 cities in Beijing-Tianjin Shayuan Project Area, including Chifeng-Yinshan agro-pastoral ecotone, Hunshandake Sandy Land, and Horqin Sandy Land. The ecosystem in this area mainly includes grassland ecosystem, farmland ecosystem, wetland ecosystem, and sand ecosystem. This area is the area closest to the desert in Beijing and Tianjin. Since the Beijing-Tianjin sandstorm source control project was launched, ecological construction in the region has been greatly strengthened, land desertification has been effectively controlled, and the control effect needs to be consolidated. The main types of land degradation in this area are wind erosion and desertification. The main prevention and control measures are to prevent and control desertification, increase afforestation and aerial seeding on slope land (sand), vigorously build windbreak and sand fixation, implement fencing and enclosure, and complete animal husbandry infrastructure.

6.1.1.3 Soil and Water Loss and Desertification Areas in the Middle and Upper Reaches of the Yellow River, Inner Mongolia

This area includes the Yellow River basin. This is a belt-shaped soil erosion area, a sand and salt area eroded by wind, water, and gravity. It involves 56 counties in the whole region and is the most seriously degraded area in the autonomous region. The ecosystem in this area mainly includes grassland ecosystem, farmland ecosystem, wetland ecosystem, and desert ecosystem. These include Mu Us Desert, Kubuqi Desert, and Ulaanbaatar Desert. The types of land degradation in this area are mainly water erosion, wind erosion, and salinization. In this area, the focus is to strengthen the protection of forest resources and the management of soil erosion and salinization of land desertification. Vigorously develop the sand industry and make full use of solar energy resources.

6.1.1.4 Inner Mongolia Alashan Shayuan District

This area belongs to the arid hinterland of the north temperate zone. Precipitation is very small (the lowest annual precipitation is only 50 mm), which is the main source of sandstorms. The ecosystem in this area is mainly desert ecosystem and oasis ecosystem. The types of land degradation are mainly wind erosion, soil coarsening, desert salinization, and oasis shrinkage. The focus of governance work in this area is to strengthen the construction of nature reserves, implement the natural forest protection project, implement the project to improve the living environment in different areas (ecological migration), increase blocking and management efforts, and strengthen the use of wind energy, solar energy, and biogas, developing high-efficiency agriculture such as four-in-one greenhouse.

6.1.2 Priority Ranking of Actions in Counties and Cities

Considering the regional land degradation status, poverty level, ecological sensitivity, and other factors, the standards are sorted and prioritized (Table 6.1), and the key areas of land degradation prevention and control actions are determined (sorted by county and city). See table below. Only flag counts with higher scores are listed:

Table 6.1 Partial flag county priority list

Flag county priority

Flag county name	Ecologically sensitive comprehensive score	Poverty comprehensive score	Overall ratings
Kerqin right wing middle flag	3.7	5	8.7
Keshiketeng Banner	3.5	5	8.5
Etuokeqian Banner	3.5	5	8.5
Wuchuan County	3.5	5	8.5
Horqin left wing flag	4.0	4	8.0
Zhabit Banner	3.0	5	8.0
Bahrain Zuoqi	3.0	5	8.0
Bahrain Right Banner	3.0	5	8.0
Naiman flag	3.0	5	8.0
Kulun Banner	3.0	5	8.0
Weng Niute Banner	3.0	5	8.0
Aohan Banner	3.0	5	8.0
Linxi County	3.0	5	8.0
Duolun County	3.0	5	8.0
Hang Jinqi	3.0	5	8.0
Arong Banner	3.7	4	7.7
Zhalantun	3.7	4	7.7
Tuquan County	3.7	4	7.7
Zalute flag	3.7	4	7.7
Zhuozi County	3.7	4	7.7
Liangcheng County	3.7	4	7.7
Positive blue flag	3.5	4	7.5
Kalaqin flag	2.5	5	7.5
Ningcheng County	2.5	5	7.5
Huade County	2.5	5	7.5
Shangdu County	2.5	5	7.5
Right rear flag	2.5	5	7.5
View right middle flag	2.5	5	7.5
Wushen Banner	2.5	5	7.5
Dalhan Maomingan United Flag	2.5	5	7.5
Siziwangqi	2.5	5	7.5
Tokto County	2.5	5	7.5
Guyang County	2.5	5	7.5

6.1.3   Biodiversity Hotspots

Biodiversity is the material basis for human survival and development. Biodiversity not only provides basic needs for food, energy, and materials. It also plays a key role in maintaining the ecological balance, stabilizing the environment, maintaining soil fertility, ensuring water quality, and adjusting the "service function" of the climate. Biodiversity includes three basic levels: ecosystem diversity, species diversity, and genetic diversity.

6.1.3.1 Assessment of Biodiversity and Habitat Sensitivity

According to the abundance of species in the habitat, i.e., the number of protected areas in the countries and autonomous regions in the region, the sensitivity of the habitat was evaluated, and the national first, second, and autonomous protected species and their distribution locations in the Inner Mongolia Autonomous Region were found or were subject to human interference. Based on the vegetation map and land use map of Inner Mongolia Autonomous Region, the species and their habitats were evaluated, and the biodiversity and habitat sensitivity map of Inner Mongolia Autonomous Region was drawn. The evaluation index is shown in the following table (Table 6.2).

Biodiversity conservation has a high global significance. The following is a description of biodiversity hotspots in Inner Mongolia based on different ecosystems.

Table 6.2 Assessment of local or habitat importance

Nature of habitat	Importance of importance
Natural	True primitive habitat>secondary habitat>artificial habitat (e.g., farmland)
The size of the habitat area	Under the same conditions, the habitat with a large area > a small area
Diversification	Areas with many communities or habitat types > single, simple areas
Rareness	Habitats with one or more rare species > habitats without rare species
Recoverability	Habitats that are easy to recover naturally > habitats that require human assistance to recover
Fragmentation	Easy-to-intact habitat > fragmented habitat
Ecological connection	Functionally related habitats > functionally independent habitats
Potential value	Through natural process or proper management, it can eventually develop into a habitat with more natural protection value than the habitat without development potential at present.
Feeding farm/breeding ground	Habitats in which species or communities breed and grow> habitats without this function
Duration of existence	Long-standing natural or seminatural habitats > newly formed habitats
Number/abundance of wildlife	Biodiversity-rich habitats>biodiversity simple habitats

6.1.3.2   Integrated Ecosystem

6.1.3.2.1   Honghuaerji *Pinus sylvestris* Forest Park

Located in the transition zone between Daxing'anling and the famous Hulun Buir forest and grassland, it is a typical forest grassland with vast forests, green grassland, and vast natural pastures. The pure forest coverage of the park is 84%. The forests in the park are well separated, with towering trees scattered among them. The highest tree age is 300–500 years. Hundreds of miles away, the shade of trees and the pine forest in the park are like a green sea. This is the most magnificent view of the pine forest.

Grassland: The Erlangen grassland is an important part of Hulun Buir grassland. In the vast and boundless grassland, a hundred flowers blossom and cattle and sheep flock together. They interpret and outline the classical and tranquil scenery outside the grassland of "sky, wilderness, wind and grass, low and looking."

Wild animals: There are 4 kinds of national protected animals and 26 kinds of national protected poultry. Other common wild animals include scorpions, rabbits, wild boars, and antelope.

Wild plants: They mainly include trees, shrubs, and surface plants. *Pinus sylvestris* var. mongolica, *Betula platyphylla*, poplar, larch, Mongolian willow, mountain stem, and thick plum are the main tree species.

6.1.3.2.2   Tumuji Nature Reserve

It is located in the transition zone between Daxing'anling and Songnen Plain. It is also a transitional zone between temperate grassland and arid grassland in China. The transitional features of geology, geomorphology, and vegetation make it unique in geographical location.

The total area of the reserve is 948.3 km², including 53.6 km² of water, 220.6 km² of wetland, 368.9 km² of grassland, 28 natural villages in 7 villages of the reserve, and a population of more than 10,000. There are 309 species of animals belonging to 71 families in the reserve, including 228 species of birds, 9 species of national protected birds, and 36 species of secondary protected birds. National birds are the biggest feature of the nature reserve. It is one of the few large-scale breeding areas in China and is known as the hometown of Dagu. There are 26 species of mammals of 10 orders, 10 families, and 5 species in the nature reserve, of which 1 species is the national secondary protected animal. There are 476 species of plants belonging to 275 genera and 79 families in the reserve, which are rich in resources. Therefore, the reserve is one of the typical areas for biodiversity research and protection of temperate grassland ecosystem and wetland ecosystem. It has great apes, cranes, and dragonflies. It is an important basis for studying the protection of birds and other resources.

6.1.3.2.3   Heilihe Nature Reserve

The reserve is located in the west of Ningcheng County, Chifeng City, Inner Mongolia, with a total area of 27,638 ha. The main protected objects are the warm temperate coniferous and broad-leaved mixed forest ecosystem and biodiversity resources represented by large natural pine forests.

Heili River Nature Reserve has a steep mountain peak 7,701,836 m above sea level.

This is a temperate deciduous broad-leaved forest area with dense forests, rich species, complex biota, and obvious vertical spectrum of vegetation. There are 3 types of plant cover and 24 groups in the reserve. There are 777 species of wild vascular plants, 540 of which are medicinal plants, accounting for 70% of the total number of vascular plants and 176 species of bryophytes. Forest resources in Heilihe Nature Reserve occupy an important position in Inner Mongolia Autonomous Region. An important natural barrier, only China's endemic species of *Pinus tabulaeformis* has a natural distribution of 4667ha. It is the largest and most growing natural pine forest in China and is very precious. There are 13 rare and endangered bryophytes in the reserve. Among them, *Astragalus membranaceus* (Huangbo) and *Actinidia arguta* are extremely endangered species in the reserve. Heilihe Nature Reserve has become an important breeding base for natural *Pinus tabulaeformis*.

There are 117 species of birds and 33 species of mammals in the reserve. It focuses on the protection of 19 species of birds and 2 species of mammals. Among them are golden sculptures of birds, shovels of chickens, leopards of mammals, black bears, etc. It is an extremely endangered species in the reserve and has attracted the attention of the whole society.

6.1.3.3   Desert Ecosystem

6.1.3.3.1   Ejina Huyanglin

Ejina poplars are one of the world's third largest poplars. *Populus euphratica* is densely distributed and has a magnificent landscape.

Hu Yang is called "Tao Lai" in Mongolian. This is a deciduous tree with delicate wood and soft leaves. It is drought-resistant and life-threatening. This is one of the rare tree species in nature. The Huyang forest area in Ejina is one of the three most completely protected areas in the world. *Populus euphratica* has existed for hundreds of years and is still thriving. It is a miracle that nature dances alone.

6.1.3.3.2   Inner Mongolia Western Ordos National Nature Reserve

The region is located on the eastern edge of the Asian-African desert. It is the transition zone between the desert steppe in western Ordos and the grassland desert in eastern Alashan. It has the ancient Mediterranean bone plants *Tetraena mongolica, Hemiptera,* cotton spine, *Ammopiptanthus mongolicus*, Pidaisy, and Mongolian almond. In Hu Yang and other places where they live, 335 species of higher plants have been identified in this area, 72 of which are endangered, accounting for 21.79% of all plant species. Seven species of wild plants are listed as national key protected species. In particular, tetraploid flowers and daytime flowers are only distributed in a small part of the protected area, which has extremely high protection and scientific research value. In addition, the reserve also preserves extremely precious ancient geographical environment. There are abundant fossils and obvious mountain strata. This is a very precious natural history book. Therefore, the establishment of this area is of great significance not only for the protection of local biodiversity but also for the study of the origin, development, and evolution of organisms, as well as the geological structure and paleogeographic environment. It is also of great practical significance to improve the living environment in desert areas of our country and to explore ways of sustainable development in desert areas.

6.1.3.3.3   Hatengtaohai

Hatengtaohai Nature Reserve is located in the northwest of Qingkou County, with Hatengtaohai Sumu and gold placer Taohai Sumu. It is 42 km wide from north to south and 53 km wide from east to west, with a total area of 123,600 ha. There are abundant wildlife resources in the reserve. A total of 302 species, 160 genera, and 53 species of seed plants were identified, including national second-class endangered plants, cotton spines, *Ammopiptanthus mongolicus*, *Cistanche deserticola*, and third-class protected plants, *Prunus mongolica*, *Haloxylon ammodendron*, and *Populus euphratica*. The area is 22,900 hectares. There are 96 species of terrestrial wild vertebrates in animal resources, including 6 species, 11 families, 27 species, 14 species, 28 families, 62 species, and 7 species of amphibians and reptiles. There are 22 species of wildlife under national protection, including the first-class and second-class protected animals. There are six kinds of national first-class protected animals, such as black scorpion, big scorpion, wave scorpion, *Capra ibex*, golden eagle, and white peony root. There are 16 kinds of secondary protected animals, such as Pan Yang and demoiselle cranes.

6.1.3.4   Grassland Ecosystem

6.1.3.4.1   Ordos Gull

Ordos Relict Gull Nature Reserve is located in Dongsheng City, Inner Mongolia, with an area of 7680 ha. It belongs to Eurasia grassland area and Asian desert area, with fragile ecology. The main vegetation is willow, valerian, and *Suaeda salsa*. Thirty-three species of wetland birds, all migratory birds, were recorded. The main birds are brown-headed gulls, swans, scorpions, swans and ducks. The nature reserve protects the natural environment and saves the remains of the endangered species gull.

6.1.3.4.2 Daxing'anling Grassland

The main protected objects are meadow grassland, typical grassland, sandy grassland, and river valley wetland ecosystem.

Xilinguole grassland is the most representative grassland of China's *Stipa grandis* (Leymus chinensis), and it is also a part of the native grassland in the Asian grassland subregion of Eurasia. The ecological environment type of the reserve is unique and has the basic characteristics of grassland biological community, which can fully reflect the structure and ecological process of typical grassland ecosystem in Inner Mongolian Plateau. At present, there are 658 species of seed plants belonging to 299 genera in 74 families, 73 species of bryophytes, and 46 species of macrofungi, of which 426 species are medicinal plants and 116 species are fine herbages. The distribution of wild animals in the reserve reflects the characteristics of Mongolian plateau. There are 33 kinds of mammals, such as antelope, wolf, and fox, and 76 kinds of birds. There are five protected wild animals, such as red-crowned crane, white stork, giant salamander, and jade belt sea eagle. The national second-class protected wild animals include 21 species such as whooper swans, grassland carvings, and antelope. This area is the largest grassland and meadow ecosystem nature reserve in our country at present. It has an important position and obvious international influence in protecting grassland biodiversity.

6.1.3.4.3 Inner Mongolia Kerqin National Nature Reserve

It is located on the right wing of Xingan League in Inner Mongolia Autonomous Region and covers an area of 126,987 ha. The main protected objects are rare birds in wetland and typical natural landscape of Horqin grassland.

This area is located in the transition zone between the low mountains and hills in the south of Daxing'anling and Horqin Sandy Land. This is a moderate temperature semiarid continental climate. The natural mosaic combination of forests, irrigation, grasslands and rivers, lakes, swamps, and other wetlands constitutes a complex and diverse habitat type, providing unique conditions for the survival and reproduction of species. There are 452 species of higher plants in 65 families in this area. One hundred sixty-seven species of 16 species of birds, including 34 national key protected animals, and breeding grounds for rare birds such as red-crowned cranes, white-necked cranes, pheasants, white pelicans, and giant salamanders. It has very important protection value.

6.1.3.5 Forest Ecosystem

6.1.3.5.1 Helan Mountain National Nature Reserve

It is located in Alashan Zuo Qi, Inner Mongolia Autonomous Region, with an area of 67,710 hectares. The main object of protection is the mountain forest ecosystem in arid and semiarid areas. This area is located on the western slope of Helan

Mountain, which is the transition zone from grassland to desert. The highest peak is 3556 m above sea level, and the relative elevation difference is 2000 m. The vertical change of vegetation in this area is obvious. Forest vegetation mainly consists of oilseed forest and *Picea crassifolia* forest. There are 511 species of higher plants, including *Ammopiptanthus mongolicus* and *Syringa oblata*, which are protected by more than 10 countries. Model samples are produced in this area. There are 32 species, 8 species, or 8 varieties of plants on the western slope of Helan. Animals in this area are also very complicated. It has representatives from North China and Meng Xin. One hundred seventy-seven species of higher animals (115 species of birds) were found, among which the national key protected animals are red deer, lynx, blue pheasant, etc. The establishment of this protected area is very important to maintain the integrity and stability of the local ecosystem.

6.1.3.5.2 Inner Mongolia Baiyinaobao Sandy Spruce Forest

Baiyinaobao National Nature Reserve is located in the northwest of Keshiketeng Banner, Chifeng City, Inner Mongolia Autonomous Region, with an area of 36,000 mu.

Baiyinaobao is located 70 km northwest of Peng Jing Town in the deep grassland. *Picea mongolica* growing here is a rare species in the world. The forest industry calls it "biological gene pool, living fossil." Baiyinaobao National Nature Reserve has a forest area of 36,000 mu, 20 km from west to east of Huanggang Liang. *Picea mongolica* is 28 m high, and this tree has a history of 100 years. The shape of this tree is like a tower, with dry red leaves and green leaves, evergreen all the year round. The forest is in the middle of saplings and has distinct levels. The river is clear and rich in flowers and plants. Spring and autumn are full of fragrance.

Baiyinaobao was the head of the five seas and the sun in the yuan dynasty. On the 13th of the lunar calendar, herdsmen come to worship. The scene is solemn and mysterious. At the foot of the mountain, spruce is built for eating, living, swimming, and entertainment. The quaint forest hut is a supplement to the traditional yurt.

6.1.3.5.3 Inner Mongolia Daqinggou National Nature Reserve

It is located in Horqin Left Wing Banner, Inner Mongolia Autonomous Region, with an area of 8183 ha. The main goal of protection is the precious broad-leaved forest in arid areas. Daqinggou is located in Horqin Sandy Land in the south of Xiliaohe River Basin. There is a precious mixed broadleaf tree forest in this ditch. Ditches are dune grasslands and sparse woodlands, in sharp contrast to the vast sandy landscape around them. The flora of this area is relatively complex, with 767 species of higher plants, among which ash, walnut, and *Gastrodia elata* are the national key protected plants.

6.1.3.5.4   Hanma Nature Reserve

Hanma Nature Reserve is located on the northwest slope of Daxing'anling, adjacent to Huzhong Nature Reserve in Heilongjiang Province and Gansu Nature Reserve in the southeast of Gansu Forestry Company. The total area is 107,348 hectares and the forest coverage rate is 60.5%.

Vegetation: The climate in this area is extremely cold, the growth period is short, the precipitation is less, and the frozen soil is widely distributed and deep. *Larix gmelinii* is cold-resistant, moisture-resistant, and thin soil-resistant and can be widely propagated, becoming a cold-warm soft coniferous forest zone in this region. In addition, there are a small number of pine and birch forests.

Animals: This area has sufficient water supply and good housing conditions. Forest mites, lichens, shrubs, and pine trees have created good living conditions for precious animals such as deer and sable. This area is home to most animals in the cold temperate coniferous forest communities in Gubei, northeastern, and Daxing'anling subregions. These wild animals have rare animals in China: moose, red deer, scorpion, etc.; Sable, raccoon and otter of carnivores; Cockroaches of the cat family, rabbits of the rabbit family and Mountain hare. Birds are mainly pheasants, grouse, Pacific birds, black pheasants, etc.

6.1.3.6   Wetland Ecosystem

6.1.3.6.1   Alu Kerqin Flag Grassland

The wetland nature reserve in Alu Kerqin Banner is located between the southern Daxing'anling, Yanbei mountain range, and the loess hilly region of Liaohe river. With a total area of nearly 310,000 ha, the land spans 10 villages and towns and 62 villages. It is China's famous Horqin Sandy Land. This important component is also a water conservation area and an important passage and resting place for birds to migrate from north to south. There are more than ten large and small lakes in the reserve, such as red-crowned crane, swan, red goose, gray goose, seagull, and blue-headed duck.

6.1.3.6.2   Huihe National Nature Reserve

Huihe National Nature Reserve is located in the southwest of Hulun Buir City, within the Ewenki Autonomous Banner, with a total area of 3464.48 km². Wetland is the main body of Huihe National Nature Reserve. There are three types of wetlands in the reserve: river type, lake type, and swamp type. It is characterized by large area combination and distribution, with a total area of 1167 km². Wetlands in the reserve play an important role in maintaining the ecological balance in the region and are ideal environments for the survival and reproduction of many rare and endangered birds.

There are 187 species of birds in 38 families in the reserve, including 9 species of protected birds such as red-crowned cranes and giant salamanders. There are 31 species in 8 families of fish, 10 species in 3 families of amphibians and reptiles, and 42 species in 15 families of mammals, of which 4 species are protected by the state. There are 344 species of plants in the reserve, belonging to 199 genera and 60 families. In recent years, driven by economic interests, lawless elements often enter the Xunhe River Basin to hunt wild animals and destroy the local biodiversity, resulting in the reduction and disappearance of some species.

6.1.3.7   Water Ecosystem

6.1.3.7.1   Darinol

The reserve is located in Keshiketengqi, Chifeng City, Inner Mongolia Autonomous Region. The geographical coordinates are 160 27′-117,000′ east longitude and 430 11′-430 27′ north latitude, with a total area of 119413.55 ha, which has special natural and historical geographical conditions. The famous inland lake ecosystem on the Inner Mongolian Plateau has 22 lakes and bubbles. The largest area in Darinol is 22,883 ha. Many lakes, rivers, swamps, and wetland meadows form a diversified wetland ecosystem, accounting for 35.8% of the total protected area. It has been listed as an important wetland in Asia. Diverse habitat conditions enrich species resources. This is the westernmost border between the world's red-crowned cranes (*Grus japonensis*) and a large number of migratory birds such as *Cygnus* and *Otis tarda*. It is an important breeding ground and migration distribution center. The study found that there are 134 species of birds belonging to 33 families and 16 orders in the reserve, 23 of which are national key protected birds. Dalinor is a big scorpion, a big swan, a *Grus vipio* and a feather. There are many kinds and quantities of waterfowl, such as Virgo human and other wild ducks. There are thousands of birds in spring and autumn, only a few thousand swans. It looks very spectacular. It is called basalt platform and lake plain of Inner Mongolian Plateau. It is the most representative chestnut grassland on the Inner Mongolian Plateau. It is also one of the best preserved natural grasslands in the world.

6.1.3.7.2   Inner Mongolia Daban Lake National Nature Reserve

It is located in Xinbaerhu Right Banner, Hulunbuir City, Inner Mongolia Autonomous Region, covering 740,000 ha. The main protection targets are wetland ecosystems and rare and endangered wildlife dominated by birds.

This area is located in the west of Daxing'anling and the east of the Mongolian plateau, with an altitude of 550–650 m. Climate is interwoven with ocean and continental climate. The natural landscape has the characteristic of transition from typical grassland to meadow grassland, which is typical and representative. Daban Lake is the largest freshwater lake in the arid region of Central Asia. The lake is extremely rich in water resources. Lakes, rivers, ponds, marshes, wetlands, and grasslands form a complex and diverse natural ecological environment. Biodiversity is located at the forefront of the grasslands in Central Asia. There are 448 species of plants and 191 species of birds in this area. Among them, 36 species of red-crowned cranes, white cranes, and black crickets belong to the national protected birds. The establishment of this reserve is of great significance to the protection of biodiversity and natural ecological environment in grassland areas of our country.

6.2 Priority Activities

6.2.1 Priority Activities to Improve the Legislative Environment of the Autonomous Region

The legislative activities carried out by the legislature of the autonomous region and the legislative environment of the autonomous region are organized within the national legal framework. However, there are also objective problems of continuous improvement, because the emergence of new situations and new problems will inevitably be reflected in the research of legislative work. The first consideration in the legislative environment is the consistency of relevant legal documents and national development policies. The policy will directly affect the legislative environment. Secondly, the public subject of legislation should have extensive participation. The construction of a harmonious society should be fully reflected in the organizational forms and mechanisms of public participation in the legislative environment. In the legislative environment, the role of "representatives of public opinion" is still lacking. The most urgent task for improving the legislative environment of the autonomous region is to rectify the legislation, weaken the trend of legislating in departments and legalizing the interests of departments, and correct the privileged behavior of industry departments. The second is to cultivate the environmental conditions for public participation in legislative activities. Laws and regulations should be the defenders of the legitimate rights and interests of every citizen. The third is to establish and perfect an effective legislative environment supervision mechanism to supervise the implementation of laws and regulations.

6.2.2 Key Activities to Improve the Autonomous Region's Development Policy Environment

The development policy environment of the autonomous region is the policy background to adapt to the continuously developing social, economic, and ecological protection and construction. Changes in the policy environment are determined by changes in the policies and lines of the autonomous region government and relevant functional departments. For example, during the "Tenth Five-Year Plan" period, the state took construction as the basic policy for social and economic development, increased investment in ecological environment protection and construction in Inner Mongolia, and issued active policies to encourage and support social forces to participate in ecological environment protection and construction. Another example is that the central government's basic policy to solve the "three rural" problems is "give more, take less, and live more." Industry supports agriculture and animal husbandry, while cities support rural and pastoral areas. Since 2004, the government has issued policy documents such as "increasing the income of farmers and herdsmen," "improving the comprehensive production capacity of agriculture and animal husbandry," "promoting the construction of a new socialist countryside," and "building a modern agriculture and animal husbandry, and safeguarding the national food ecology." It has laid a foundation for the economic development of rural and pastoral areas, increased the income of farmers and herdsmen, and improved their living standards and quality of life. The implementation of central and autonomous government policies depends on the impact of the policy environment. The first is to promote policies so that they can be implemented at the grassroots level. The first activity to improve the development policy environment of the autonomous region is to clean up the documents. In recent years, in order to meet the needs of the autonomous region's social and economic development, the autonomous region government has organized several times to clean up documents, cancelled more than 2000 documents and management regulations that are not suitable for development, relaxed the relevant examination and approval authority, and effectively promoted the market development needs. Cleaning up documents is an effective means to standardize and improve the autonomous region's development policy environment. Therefore, the best choice to optimize the autonomous region's policy environment is to regularly clean up relevant policy documents and continuously improve and formulate corresponding rules and regulations.

6.2.3 Priority Activities for Capacity Building

Strengthening the institutional capacity building of autonomous regions and local government agencies, carrying out integrated ecosystem management, and controlling land degradation are top priorities.

To carry out activities related to the control of land degradation and integrated ecological management, autonomous region institutions and local policy organizations at all levels need the cooperation and participation of relevant institutions, units, and individuals and establish multiple institutions engaged in integrated ecological management within relevant organizations. Responsibilities and obligations are shared by multiple departments and members.

In the organization and implementation of related activities, integrated ecosystem management activities are carried out according to the responsibilities of organizations at all levels, including autonomous region institutions, governments at all levels, government leaders, and leaders of competent departments.

Scientific research units, technical service organizations, and various economic organizations that play an active role in the integrated ecosystems management should be given corresponding powers and responsibilities for the management of integrated ecosystems and play their roles.

Organizing study and training is a priority activity needed to improve the capacity of autonomous regions and local government agencies. Through research, participants and managers have a better understanding of IEM concept. Through the organization and practice of integrated ecosystem management and control activities related to land degradation, ecological protection and construction can be effectively integrated and sustainable.

The establishment of the autonomous region cultivated land quality management information system and early warning system. In accordance with the requirements of protecting cultivated land and improving grain production capacity, we will strengthen the investigation of cultivated land quality, environmental monitoring, and construction work. Through the demonstration of the project, the scope of the investigation of cultivated land fertility will be further expanded; the classification evaluation of cultivated land, soil fertility, and environmental monitoring will be carried out in an all-round way; changes in cultivated land quality will be grasped in a timely manner; the environmental quality, production capacity, and distribution of cultivated land in the autonomous region will be understood; and an information system and early warning system for the management of cultivated land quality in the autonomous region will be established (Table 6.3).

6.2.4 Screening and Promotion of Best Practices Priority Activities

From the nearly 100 measures for land degradation prevention recommended by various industries, we have selected the following technical measures through expert evaluation (Table 6.4):

6.2.5 Priority Activities for Poverty Alleviation

Increase labor transfer training efforts. Reduce the labor force on the existing land, and promote the transfer to cities and towns through the "sunshine project" implemented by the government to reduce the pressure on the land.

Implementation of immigration and resettlement. Most of the existing poor people live in areas with poor ecological environment. In order to reduce the damage of this part of the population to the land and prevent the occurrence of soil erosion, it is neces-

Table 6.3 Priority activities for capacity building in Inner Mongolia Autonomous Region

Serial number	Name	Department	Object	Content
1	IEM management	Member unit	Mechanism	Mission, philosophy, partnership, dynamic management
2	Laws and policies to combat land degradation	Member unit	Mechanism	Develop and improve relevant laws and policies
3	GEF Information Management System	Forestry	Mechanism	
4	Evaluation tools and methods for GEF projects	Member unit	Personal	
5	Interdepartmental coordination mechanism for combating land degradation	Development and reform commission	Mechanism	
6	Incentives for combating land degradation demonstration communities	Member unit	Community	
7	The application of participatory concepts in GEF	Member unit	Personal	
8	The application of participatory methods in community development	Member unit	Community	
9	Integrated use of funds in GEF projects	Financial	Mechanism	
10	Application of Logical Framework Method (LFA) in GEF project	Member unit	Personal	
11	Ecological environment monitoring system	Environmental protection	Mechanism	
12	Assessing the framework of the land degradation indicator system	Water conservation, environmental protection	Mechanism	Determine the evaluation system and indicator system
13	GEF project economic analysis and methods of priority activities	Member unit	Personal	Mission, philosophy, partnership, dynamic management
14	Method of biodiversity hotspot analysis	Forestry, environmental protection	Personal	Develop and improve relevant laws and policies

Table 6.4 Functions of best practices in land degradation prevention

Strategic option	Land degradation consequences					
	Cultivated land degradation	Grassland degradation	Forest degradation	Water resource degradation	Biodiversity degradation	Poor
keep status quo	▓	▓	▓	▓	▓	▓
1. Sand bio-economic circle	▓	▓	▓			▓
2. Increase the application of farmyard manure	▓	▓				▓
3. Dry farming technology	▓					▓
4. Returning farmland to forests (grass)	▓	▓	▓			▓
5. Fenced grassland		▓			▓	
6. Artificial grass		▓				
7. Black land management	▓	▓				
8. Grassland rodent		▓			▓	
9. Seasonal grazing		▓				
10. House feeding		▓				▓
11. Adjust the herd structure		▓				▓
12. Artificial fore station			▓			
13. Hills for affo restation		▓	▓			
14. Aerial seeding			▓			
15. Windproof sand	▓		▓			
16. Protecting endangered wildlife				▓	▓	

(continued)

Table 6.4 (continued)

17. Wetland protection				●	●		
18. Small waters hed management	●			●			
19. Ecological restoration				●	●		
20. Farmland flood control dam	●			●			
21. Supplementary irrigation	●						●
22. Disaster warning	●	●		●			●
23. Climate Change Response Engineering	●	●	●				
24. Artificial wea ther engineering	●	●	●				
25. Land Resour ces Survey and Evaluation	●	●	●				
26. Land develop ment and reclam ation	●	●	●	●			
27. Agricultural l and classification and grading	●						
28. Ex situ poverty alleviation and development	●	●					●
29. Village prom otion	●	●					
30. Labor transfer training	●	●					●
31. Industrializati on poverty allev iation							●
32. Environmenta l protection capa city building				●	●		
33. Solar power		●					
34. Solar cooker	●	●					

(continued)

Table 6.4   (continued)

35. Biogas pool						
36. Integrated and coordinated partnership						
37. Raising special funds						
38. Technical trai ning						

sary to implement resettlement as soon as possible and to move this part of the population to the edge of cities and towns with better conditions, promoting urbanization.

The conversion of crop straw is implemented in agricultural areas. Dispose of the remaining straw in agricultural areas, raise livestock, and return the soil to the soil to improve the soil, thus reducing soil erosion. Through the introduction of agriculture and grazing, the southern feeding method, the surplus pasture in the agricultural areas will be transported to the pastoral areas, and the livestock in the pastoral areas will be fed. The reproduction rate of livestock will be shortened, the feeding period of livestock will be shortened, and the number of livestock will be reduced. Through years of checking, the pressure on livestock and grassland will be reduced and soil erosion will be prevented.

Build rain rafts. The shortage of water resources is an important factor that restricts the social and economic development of Inner Mongolia and also restricts the development of agriculture and animal husbandry in poor areas. In summer, rainwater feeds animals in water. In the spring of the following year, water is used for afforestation, so as to maintain water and soil and reduce the impact of soil erosion.

Planting shrubs, sand, salt and sea buckthorn can prevent further expansion of sandstorm and desertification area. On the other hand, local farmers and herdsmen can achieve their income through the flatness of *Salix*.

Back to the ranch. The solution to the problem of destroying cultivated land is grazing and returning farmland to grass. Farmers and herdsmen have solved the problem of food and clothing through state grain subsidies and planting grass and grass to alleviate and restore land degradation.

Increase the promotion of new irrigation equipment, change irrigation methods, and implement water-saving irrigation. The promotion of new irrigation equipment, the implementation of water-saving irrigation, and the combination of sprinkler irrigation and drip irrigation in areas where conditions permit are conducive to soil improvement and prevention of land degradation.

Through the introduction of new animal husbandry varieties, the original varieties have been improved. The introduction of new breeds has improved the reproduction rate of sheep, accelerated the slaughter rate of mutton sheep, and increased the unit meat yield of sheep, thus increasing the income of farmers and herdsmen, shortening the breeding cycle of farmers and herdsmen, reducing the pressure on grasslands, and slowing down the process of land degradation.

Construction of biogas digesters in rural areas. In combination with the progress in the construction of rural biogas digesters in Wenniu Special Banner and other areas in Inner Mongolia, Inner Mongolia has vigorously promoted the construction of biogas digesters, using human and animal waste as raw materials.

Soil fertility. In view of the decline in natural fertility of cultivated land in the whole region, we are actively exploring scientific farming, planting, and fertilization systems. At the same time of adding organic fertilizer, according to the fertilizer demand of crops, soil fertility performance and fertilizer efficiency, determine the soil nutrient content, scientifically determine the proportion and quantity of NPK and trace elements, develop formula fertilization technology, improve soil fertility, improve soil structure, fertility and fertility, improve cultivated land quality and production capacity, and take the road of sustainable utilization.

6.2.6 Priority Should Be Given to Activities to Protect and Improve Biodiversity in the Autonomous Region

6.2.6.1 To Investigate the Wild and Endangered Animal and Plant Resources in the Whole Region

In recent decades, due to the influence and interference of human activities, wildlife resources have been more seriously damaged. In order to understand the current situation of endangered wildlife in the region as soon as possible, it is urgent to investigate the wildlife resources in the region and take targeted protection measures.

6.2.6.2 Strengthening the Construction of Nature Reserves

The construction of nature reserves is an important means to protect rare species resources, maintain genetic diversity, protect the original natural landscape and various habitat types, and protect natural resources and wild animals from extinction and reproduction. According to the needs of biodiversity protection in Inner Mongolia, the nature reserves that need priority protection are as follows (Table 6.5).

6.2.6.3 Establish a Biodiversity Monitoring System

To carry out biodiversity monitoring and scientific research is a necessary step for the protection and sustainable use of biological resources. Through monitoring, we can understand the changes of population growth and distribution, understand the endangered factors of biological population, and study the methods to solve the problem of ambulance flow of endangered population. The state provides scientific basis for the protection, management, and utilization of policies. In particular,

Table 6.5 Nature reserves priority construction in Inner Mongolia

Serial number	Name	Position	Area(ha)	Protected object
1	Darinor Nature Reserve	Dalhan Sumu, Dali town	119,414	Rare birds
2	Western Ordos National Nature Reserve	Ordos Eqi	539,000	Rare plants such as Sihemu\ diversified ecosystems\ grassland to desert transition vegetation belt
		Wuhai City	16,898	
3	Inner Mongolia Autonomous Region Tumuji National Nature Reserve	Inner Mongolia Zhabite Banner Tumuji Sumu	94,830	Daxie
4	Huihe National Nature Reserve	Ewenki banner	346,848	Rare birds, wetlands, grasslands
5	Xilin Gol Prairie National Nature Reserve	Xilinhot city	580,000	Grassland
6	Da Lat Lake National Nature Reserve	Manzhouli	743,600	Birds, wetlands
7	Saihanwula Nature Reserve	Bahrain Right Banner, Chifeng City, Inner Mongolia	100,400	Forests, grasslands, ecosystems, and biodiversity
8	Ejinaqi Qidaoqiao Huyanglin Nature Reserve	Eqida Town	26,253	Populus
9	Aru Kerqin Nature Reserve	Aqi East	136,800	Rare birds, wetlands, Horqin Sandy Land, grasslands

fixed-point location monitoring stations should be established in biodiversity-sensitive areas for long-term location monitoring.

6.2.6.4 Strengthen Law Enforcement and Publicity and Education on Biodiversity Conservation

The state has promulgated laws and regulations for the protection and management of biodiversity in the Law of the People's Republic of China on the Protection of Wild Animals, the Forestry Law, and the Grassland Law. However, in some areas, for the sake of economic benefits, there are still laws that do not conform to the law. Blind development poses a serious threat to biodiversity. Therefore, it is necessary to strengthen the enforcement of biodiversity conservation and strictly implement the State Council's regulations prohibiting the development of nature reserves and areas with special conservation value. Development and construction projects and decisions that have a significant impact on the environment must strictly implement the environmental impact assessment system.

Biodiversity conservation needs the support and attention of all social strata. The importance of biodiversity to human survival and life must be raised through publicity and education of laws and regulations, so as to raise the awareness of biodiversity and protection of the whole society.

6.2.7 Priority Activities for Disaster Reduction and Prevention Priority Activities Required for Reducing Geological Disasters (Natural Disasters)

6.2.7.1 Compilation of Geological Disaster Prevention Plan in Inner Mongolia Autonomous Region

The compilation of the "Geological Disaster Prevention Plan of Inner Mongolia Autonomous Region" is a programmatic document guiding the investigation, prevention, management, and research of geological disasters in the autonomous region. The overall objective of the prevention and control of geological disasters in the autonomous region is to establish a relatively perfect legal system that meets the requirements of the socialist market economy within 15 years and strictly control the occurrence of man-made geological disasters. To strengthen the basic investigation work, establish and gradually improve the prevention system of geological disaster monitoring and forecasting and group control teams, on the basis of a basic grasp of the distribution of geological disasters in the autonomous region and the degree of harm, to mobilize the enthusiasm of all aspects and to strengthen the management of geological disasters. Major geological disasters and major hazard sources have been basically rectified. Through the implementation of planning and management, the prevention and control of geological disasters have changed from a scattered and passive emergency to an organized, professional, forward-looking, and predictive work, combining regions with priorities and reality with the long term. The combination of prevention and control, disaster reduction, and economic development has become an integral part of the autonomous region's social and economic development planning, greatly reducing the occurrence and loss of geological disasters in the autonomous region.

6.2.7.2 To Carry Out More Extensive and In-Depth Geological Disaster Investigation, Assessment, and Zoning Work

On the basis of the existing work, continue to do a good job in the investigation, evaluation, and zoning of basic geological disasters in the whole region and the region; geological hazard exploration and evaluation in key areas (cities, large rivers, railway lines, important energy bases, and major projects); investigation and evaluation of geological disaster-prone areas; and risk assessment of geological disasters in construction land.

No matter which aspect of the investigation and evaluation work, we should pay attention to the application of new technologies and methods and continuously expand and deepen the investigation content. While paying attention to the investigation and analysis of the natural characteristics of geological disasters, special attention should be paid to the investigation and study of the socioeconomic attributes of geological disasters. Pay attention to the direct hazards of geological disasters, and strengthen the investigation and analysis of the profound hazards of geological disasters and their impact on the sustainable development of social economy. At the same time, through the investigation and analysis of the natural dynamic process and direct causes of geological disasters, the research on the human factors and socioeconomic background of geological disasters should be strengthened. When investigating the history and current situation of geological disasters, strengthen dynamic analysis and prediction evaluation. At the same time of analysis, quantitative analysis and evaluation should be added. The investigation and assessment of geological disasters in the autonomous region will be comprehensively upgraded to a new stage.

6.2.7.3 To Strengthen the Construction of Modern Geological Disaster Early Warning and Forecasting Information System, Improve the Social System of Prevention and Treatment

On the basis of the existing work, further organize social forces, develop and improve the group monitoring and prevention system for geological disasters, expand the coverage area of the group monitoring and prevention system, and improve the effectiveness of the group monitoring and prevention system. Incorporate the geological disaster group monitoring and prevention system into the geological disaster early warning system, and comprehensively improve the level of geological disaster early warning and prevention. Establish an autonomous regional geological disaster information center. Establish a database of geological disasters and prevention and control, and on this basis, apply GIS technology to establish a geological disaster information system covering the whole region. Provide information services and technical method support for timely provision of geological disaster information, geological disaster assessment, dynamic zoning, and prevention and control, and provide channels for exchanges and cooperation at home and abroad.

6.2.7.4 Strengthen Scientific Research and Legalization to Provide Theoretical, Technical, and Legal Guarantees for the Prevention and Control of Geological Disasters

Mobilize the scientific research strength of universities and research institutes; learn traditional theories and methods; widely apply new theories, new technologies, and new methods; strengthen the spatial and temporal distribution of geological disasters in the autonomous region; form conditions; and become a disaster-prone mechanism. The relationship between geological disasters and social and economic development; geological hazard exploration and evaluation methods; monitoring and forecasting technology of geological disasters; geological disaster management methods and techniques; scientific research such as geological disaster mitigation management has provided theoretical and technical support for the prevention and control of geological disasters.

We will further improve laws, regulations, and technical standards for the prevention and control of geological disasters, the investigation and assessment of geological disasters, the prediction of geological disasters, and the investigation, design, construction, and acceptance of geological disaster prevention and control projects and establish as soon as possible an autonomous region geological disaster prevention and control system composed of national laws and local laws and regulations. To study the management regulation system and the standard system composed of national standards, industry standards, and local standards, so as to realize the legalization and standardization of geological disaster prevention and control work. Through various forms, we will strengthen the publicity and education of geological disasters; reduce disasters; raise the environmental awareness and disaster reduction awareness of government departments, enterprises, and the people; popularize the knowledge of prevention and control of geological disasters; and make geological disasters the conscious action of the whole society.

6.2.8 Striving for Investment Priority Activities

Due to differences in laws, norms, and government systems, the political and legal scope for coordinating priority activities is often at the national level, and many private sector and private capital activities are influenced by national economic, monetary, trade, and social security regulations. In fact, the central government plays a special role in providing a legal framework to create an environment conducive to the formation of cooperative relations between the private sector, civil society, and governments at all levels. Cooperation between government and civil society can greatly improve development results and play a key role in environmental governance. The main coordination means are:

- Create a good investment environment, strengthen government management, and provide infrastructure.
- Effectively utilize resources, ensure rational use of external aid, and enhance the government's sense of responsibility and credibility.
- Increase opportunities to participate in private capital and civil society, improve their voice, increase the inclusiveness of the system, and unleash creative social potential.

Chapter 7
Application of PPP Model in the Prevention and Control of Land Degradation in Our Region

7.1 Laws and Regulations on the Prevention and Control of Land Degradation

7.1.1 National Level

Since the reform and opening up, the state has implemented laws and regulations such as the "Soil and Water Conservation Law," "Regulations on the Implementation of Soil and Water Conservation," "Sand Control Law," "Environmental Impact Assessment Law," and "Regulations on the Implementation of Forest Law." It has revised and improved the "Grassland Law," formulated and promulgated a series of policies and guidelines on environmental management and protection that benefit farmers, and established legal and policy systems such as the "Soil and Water Conservation Law," "Forest Law," and "Grassland Law." These laws and regulations clearly define the protection of land resources, maintain the potential of land production, improve the protection of land productivity, and effectively ensure the smooth prevention and control of land degradation.

The current "Land Administration Law of the People's Republic of China" was revised and adopted at the 11th meeting of the Standing Committee of the 10th National People's Congress on August 28, 2004, and will come into force from the date of adoption by the National People's Congress. Articles 35 and 74, respectively, stipulate the prevention and control of land desertification from the perspective of farmland utilization: "People's governments at all levels shall take measures to maintain irrigation and drainage engineering facilities, improve soil, strengthen foundation strength, improve soil, strengthen strength, and prevent land desertification, salinization, soil erosion and land pollution." "in violation of the provisions of this law, unauthorized occupation of cultivated land to build kilns, tombs or houses on cultivated land, sand digging, quarrying, mining, earth borrowing, etc. If the cultivation conditions are damaged or the land is desertified or salinized due to land development, the land administrative department of the people's government at or

© Science Press & Springer Nature Singapore Pte Ltd. 2020
Z. Meng et al., *Public Private Partnership for Desertification Control in Inner Mongolia*, https://doi.org/10.1007/978-981-13-7499-9_7

above the county level shall order it to make corrections or investigate criminal responsibility according to law."

To sum up, at present China has enacted nearly 20 laws to combat desertification, covering natural resources and ecological environment protection laws, such as the State Council's Decision on Further Strengthening the Work of Combating Desertification, the Measures for Profit Management, the Grassland Law of the People's Republic of China, the Law of the People's Republic of China on Combating Desertification, the Law of the People's Republic of China on Environmental Protection, the Forest Law of the People's Republic of China, and the Law of the People's Republic of China on Soil and Water Conservation. National and local governments at all levels have formulated and promulgated a series of supporting laws and regulations. The newly revised "Land Administration Law" and "Criminal Code" have added provisions to combat desertification, basically forming a system of environmental laws and regulations that combines special laws with relevant laws, national laws, and local laws.

At the same time, the Chinese government has incorporated the prevention and control of desertification into its national economic and social development plans. It has successively formulated such important documents as China's Agenda 21, China's Agenda 21 Environmental Protection Agenda, China's Agenda 21 Forestry Action Plan, and the National Ecological Environment Plan. It has implemented the coordinated development strategy of environment and population aimed at achieving economic, social, resource, and sustainable development, thus laying a legal foundation for the prevention and control of desertification.

7.1.2 Autonomous Region

The relevant laws currently enacted in Inner Mongolia Autonomous Region include the implementation of the Law of the People's Republic of China on Measures to Combat Desertification, as well as legislation related to the prevention and control of Inner Mongolia Autonomous Region, such as the Inner Mongolia Autonomous Region Grassland Management Regulations and the Inner Mongolia Autonomous Region Grassland Management Regulations, Regulations on Geological Environment Protection, etc.

On July 31, 2004, the tenth meeting of the Standing Committee of the tenth People's Congress of Inner Mongolia Autonomous Region passed the "Measures of Inner Mongolia Autonomous Region for Implementing the Law of the People's Republic of China on Combating Desertification." It was implemented on September 1, 2004. Local regulations have six chapters and 34 sections. The first chapter of the "General Principles" clearly stipulates that the legislative basis, implementation scope, desertification control planning, objectives, and responsibility assessment of desertification control are limited, the rights and obligations of units and individuals are separately stipulated in the process of desertification control, and the rights and obligations of desertification control are necessary. The second chapter "Defense

Prevention and Control Plan" specifically stipulates the following: to be responsible for the compilation of the planning subject and to stipulate different treatment schemes for the forbidden protected areas, restoration protected areas, and governance areas. Chapter III, "Prevention of Land Desertification," specifies monitoring, statistics, and analysis of land desertification in the whole region, as well as obligations of protected areas. It is required to abide by relevant laws and regulations and take necessary protective measures in the fields of cultivation, aquaculture, processing, mining, and other activities in order to prevent the land desertification from worsening. It also made detailed provisions on strengthening grassland management and construction, strengthening water resources allocation and management, developing water-saving agriculture and animal husbandry industries, and preventing over-exploitation and utilization of water resources, vegetation degradation, and land desertification. Chapter IV, "Desertification Land Control," specifically stipulates the following: the main body of profit-making desertification control, incentive measures, guaranteed methods, and the use of desertification control funds. Chapter V of Zhe's "Legal Liability" stipulates who should be punished for violating the provisions of these Measures.

According to the description of the content, the author has refined the key terms and specific viewpoints: Article 6 of the Measures of Inner Mongolia Autonomous Region for Implementing the Law of the People's Republic of China on Combating Desertification, "Units and individuals that use land have the obligation to prevent land desertification, while units and individuals that use land desertification have the responsibility to control land desertification." Article 7: "No unit or individual may infringe upon the lawful rights and interests of desertification control." In addition, in Article 11: "The planning of desertification prevention and control shall be carried out according to the natural conditions such as the geographical location of the desertified land, land type, vegetation, climate and water resources conditions, degree of desertification of the land, etc. Ecological and economic functions, classified protection, comprehensive management and rational utilization of desertified land." It reflects the internal requirements of comprehensive ecological management and conforms to the guiding ideology of adjusting measures to local conditions. Chapter III of the Measures sets forth the obligations and prohibitions to combat desertification, such as Article 16, paragraph 1: "Any activities that destroy vegetation in prohibited protected areas are prohibited." Article 17 stipulates: "During the restoration of protected areas, felling of trees and other plants and reclamation activities are prohibited. Approved by the forestry administrative department of the people's government at or above the county level, appropriate activities such as tending, rejuvenation and replanting can be carried out to improve the ecological function of the protected areas. Article 24 of Chapter IV of the Measures stipulates: "Encourage, support, and guide units and individuals to engage in profit-making sand control activities. Units and individuals that do not have the right to land ownership or use shall obtain the right to land use or contractual management rights in accordance with the law if they engage in profit-making activities. Units and individuals engaged in profit-making desertification control activities on state-owned desertified land shall sign a control agreement with the owner or user of the desertified land to obtain the right to use the desertified land or the right to contracted management in accordance with the law. Units and individuals engaged in profit-making desertification control activities on collectively owned and uncultivated desertified land shall sign control agreements with collective economic organizations to obtain the right to contracted management of desertified land in accordance with the law. Units and individuals engaged in profit-making desertification control activities on collectively owned desertified land contracted to households shall sign a control agreement with the contractor within the contract period to obtain the right to contracted management of desertified land in accordance with the law. The term of governance and other rights and obligations of both parties shall be stipulated in the agreement of the governance agreement. This paper summarizes in detail the specific methods and approaches of profit-making desertification control activities, which provide a guarantee for profit-making desertification control. Finally, Chap. V of the Measures stipulates the determination of legal liability, such as Article 31: "In violation of the provisions of these Measures, the felling of trees, other plants, and reclamation activities shall be carried out by the people's governments at or above the county level in the restoration of protected areas. Government forestry or other relevant administrative departments shall, in accordance with their respective functions and duties, order them to stop illegal activities, confiscate their illegal income, and may impose a fine of not less than one time but not more than three times their illegal income. If there is no illegal income, a fine of less than 10,000 yuan may be imposed. Criminal responsibility shall be investigated according to law. Specific legal responsibilities will help combat desertification and provide legal protection for desertification control.

"Inner Mongolia Autonomous Region Grassland Management Regulations" is not only the first law in our country but also the law for the protection, construction, and rational use of grasslands in Inner Mongolia Autonomous Region. Since then, Inner Mongolia Autonomous Region has made two major amendments to the "Grassland Management Regulations of Inner Mongolia Autonomous Region" in 1991 and 2004, respectively. In 1998, the people's congress of the autonomous region formulated and promulgated the regulations on basic grassland protection in Inner Mongolia Autonomous Region according to the needs of grassland construction and protection in the autonomous region. In 1998, 1999, and 2000, the people's government of the autonomous region also promulgated and implemented the "detailed rules for the implementation of grassland management in inner Mongolia autonomous region," "measures for the transfer of the right to contracted management of grasslands in inner Mongolia autonomous region," and "regulations on grassland industry in Inner Mongolia Autonomous Region," respectively. The provisions of the three government regulations form a legal system of "one law, two cases, and three regulations" for grassland legislation in the autonomous region. In 2005, according to the new Inner Mongolia Autonomous Region Grassland Management Regulations, the Autonomous Region People's Government revised the regulations and merged the original three government regulations into the new Inner Mongolia Autonomous Region Grassland Management Regulations. Since then, Inner Mongolia grassland legislation has formed a "one law, one case, one

rule" system. In addition to the construction of laws and regulations, the party committees and governments of the autonomous region have also issued more than ten
normative documents to meet the needs of grassland management, construction,
protection, and grassland law enforcement.

Especially in 2007, the people's government of the autonomous region issued a
special notice on grassland supervision and management in Inner Mongolia
Autonomous Region, strengthening the supervision and management of grasslands.
Grassland ecological construction supervises desertification control, and local laws
and regulations will be carefully managed to improve operability. For example, in
the Regulations and Detailed Rules, the types and calculation methods of compensation for land expropriation are specified, and a license system is required for all
kinds of grassland occupation activities. In accordance with the provisions of these
rules, the autonomous region has uniformly printed licenses for collecting soil from
grasslands for mining, quarrying, mining mineral resources activities, grassland
wildlife purchase licenses, etc., and has fully implemented the license system.

Since 2004, China has issued three policies to support the development of agriculture and rural economy. Increased efforts to implement the benefiting-farmers
policy, mainly through the reduction of agricultural tax and increased investment in
agriculture, improving the agricultural comprehensive production capacity; through
the grain direct subsidy policy to mobilize the enthusiasm of farmers; Through the
guidance of grain subsidies to accelerate the pace of quality update; Through the
purchase of agricultural machinery subsidies, accelerate renewal of agricultural
machinery, improve the level of agricultural mechanization. In addition, the government has issued two documents, *Opinions on Accelerating the Economic
Development of Rural Areas and Trying to Increase the Income of Farmers and
Herdsmen in Rural Areas* and *Opinions on Further Promoting the Industrialization
of Agriculture and Animal Husbandry*, and put forward specific policy measures:

First, the exemption of agricultural tax. 2005 all exempt from agricultural taxes
and other agricultural specialty product taxes in addition to tobacco; agricultural tax
relief amounted to 580 million yuan, to benefit from the scope of the "Regulations
on Geological Environment Protection in Inner Mongolia Autonomous Region"
adopted at the fourth meeting of the Standing Committee of the Tenth People's
Congress of Inner Mongolia Autonomous Region in 2003. The regulations clearly
stipulate legal responsibilities, such as Article 29 of the regulations: "If the exploration and exploitation of mineral resources causes damage to the geological environment or induces geological disasters, the administrative department of land and
resources of the people's government at or above the county level shall order the
restoration and treatment within a time limit; If it fails to do so within the time limit,
it shall be fined 1 million yuan to 5 million yuan. If the circumstances are serious, the
administrative department of land and resources shall revoke its exploration license
or mining license; If a crime is constituted, criminal responsibility shall be investigated according to law. If the construction of the project causes damage to the geological environment or geological disasters, the administrative department of land
and resources of the people's government at or above the county level shall order the
restoration and treatment within a time limit; If no correction is made within the time

limit, it shall be reported to the people's government at the same level for compulsory treatment and punishment. A fine of not less than 10,000 yuan but not more than 50,000 yuan; If a crime is constituted, criminal responsibility shall be investigated according to law." Thirtieth in violation of the provisions of this ordinance, one of the following acts, the administrative department of land and resources of the people's government at or above the county level, shall be ordered to correct or ordered to stop the illegal act and impose a fine of 5000–30,000 yuan. However, according to the current regulations, the special problems of some flag cities (counties) have still not been well solved, resulting in the lack of clarity and flexibility in the applicable laws in some places. However, there are some specific operating methods in some normative legal documents, but there are also some specific operating methods in some normative legal documents, such as the Notice of the General Office of the Inner Mongolia Autonomous Region People's Government on Amending the General Plan for Land Use and Protection of Cultivated Land and Grassland and the Notice of the General Office of the Inner Mongolia Autonomous Region People's Government on Forwarding the Opinions of the Autonomous Region, Notice of the Environmental Protection Bureau and other departments on further improving the construction and management of nature reserves, Notice of the General Office of the Inner Mongolia Autonomous Region People's Government on afforestation in rainy seasons, and Notice of the General Office of the Inner Mongolia Autonomous Region People's Government on coordinating the investigation and inspection work of the desertification dynamic analysis expert group in the whole region.

7.2 The Autonomous Region Level Relevant Policy

There are 3.5 million farmers and 11.78 million farmers and herdsmen in Inner Mongolia. First, direct subsidies to grain producers reached 580 million yuan. In 2006, Inner Mongolia was exempted from agricultural tax, with a total amount of 1.276 billion yuan.

The second is subsidies for the purchase of fine grain crops and agricultural machinery. In 2004 and 2005, China invested a total of 1.165 billion yuan in seed subsidies, direct subsidies for grain producers, and subsidies for agricultural machinery in Inner Mongolia. The autonomous region's finance also increased subsidies and support. From 2000 to 2005, the state and the autonomous region invested 1.2 billion yuan in the industrialization of agriculture and animal husbandry, effectively raising the level of economic development of agriculture and animal husbandry.

Third, we will increase investment in irrigation and water conservancy construction, intensify the construction of agricultural infrastructure and the protection of basic farmland, and start the implementation of comprehensive agricultural land development and management projects.

Fourth, organize the implementation of key projects such as the national high-quality grain industry project, grain project, and geotechnical test to steadily develop grain production.

Fifth, continue to optimize the planting structure, steadily expand the agricultural area, and actively explore the development of disaster-avoidance agriculture to improve the overall efficiency of planting industry.

Sixth, we will comprehensively improve the level of industrialization of agriculture and animal husbandry and continuously increase the proportion of processing of agricultural and animal husbandry products in the farming and industrial chain. Encourage the development of advantageous industries and leading industries. While gradually realizing the development of local agriculture and animal husbandry with "one village, one product" and "one township, one industry," Inner Mongolia will actively support the export of labor-intensive products such as horticultural products, aquatic products, and livestock products with comparative advantages and will raise the level of opening up agriculture and animal husbandry to the outside world.

7.2.1   The Legal System for the Prevention and Control of Desertification

The design of the legal system of desertification prevention and control should first be based on the basic theories of legislation, such as the theory of sustainable development, the principle of minimum factors, and the theory of property rights of natural resources. Second, it should reflect the legislative purposes and principles. Therefore, based on the existing legal system, the author puts forward the property right system, incentive system, economic and social impact assessment system, and regional functional division system. This article will not discuss the existing legal systems such as approval system; monitoring, reporting, and publishing system; sand vegetation protection system; and closed protected areas system.

7.2.2   Property Right System

Property right is an arrangement of the market system. A clear definition of property rights is the basis for the occurrence of transaction behavior and the effective operation of market mechanism. The effective operation of market can achieve the greatest effect. Clearly defining the rights and responsibilities of property rights rules can prevent the occurrence of external non-economy and realize the internalization of external environment. Clarifying property rights has the following meanings: First, clarifying private property rights can avoid direct government intervention on enterprises and individuals and can avoid single choice of individuals in society and disperse decision-making, thus realizing market mechanism to control economic externalities and save processing costs. Second, a clear definition of property rights can enhance people's awareness of rights. Third, clear property rights can promote the optimal allocation of resources. Fourth, clear property rights can make people have a clear expectation, which is not only conducive to the settlement of infringement disputes but also more important is to prevent infringement, that is, under the condition of protecting property rights, people can obtain a firmer return. This sense of security encourages people to invest, thus promoting economic development. When property rights are violated, especially when income rights and transfer rights are restricted, people cannot determine the return, and the incentive function of property rights will be discounted or nonexistent.

A single passive situation makes more subjects participate in desert control. It is necessary to introduce and implement a system to protect private legal property rights so that private property rights are not violated. In the process of combating desertification, the government-led governance model sometimes leads the government to violate private property rights, that is, the government itself violates property rights. For example, as early as 1991, the State Council issued the Notice on Policy Measures to Combat Desertification, the State Administration of Taxation issued the Notice on Tax Concessions for Combating Desertification and Reasonable Exploitation and Utilization of Desert Resources, and local governments at all levels also issued a series of preferential policies such as "Who Developed, Who Controlled, Who Beneficiated" and "Inheritance of Desert Land Use Rights." In afforestation, they used all the forest policies of "Who Made Who Has." The introduction of these policies made some marginal hills and wasteland, after years of contracting, leasing, and purchasing, put on green clothes. However, with the deterioration of the ecological environment and the enhancement of social ecological awareness, China has adopted a policy of restricting private individuals from exercising their legitimate rights. Shi Guangyin and Niu Qinyu cause poverty due to desertification. They contracted the land in the 1980s, and now the land has become 10 million mu of forest land. They should have recovered their investment and obtained profits. However, according to the current policy, the desert they contracted was an act before 1999. According to the Decision of the State Council on Further Perfecting Policies and Measures for Returning Farmland to Forest (Guo Fa (2002) No.10), it does not fall within the scope of "returning farmland to forest" or the cost of enjoying seedlings for afforestation on barren hills. Moreover, the state has also adopted the "forest cutting policy," making the investment cost unable to be recovered. Therefore, they can only continue to operate and invest, but they cannot properly recover costs and profits in order to continue to expand renewable property rights. The government's dishonesty is an infringement on the legitimate rights of private individuals, which has caused a very bad influence and greatly dampened the enthusiasm of private forces to participate in social welfare undertakings.

How to perfect the property right system of natural resources in the process of combating desertification? Firstly, the ownership and management rights should be separated, so as to form an economic contractual relationship between the owner (the owner of natural resources is mainly the state) and the operator. A market-oriented property right system should be established. On the premise of balancing the public interests of the owner and the user, natural resources with relatively clear boundaries should be allocated or auctioned to different property rights subjects, including the state, local governments, enterprises, and individuals, according to their uses, publicness, and externalities of enterprises. For natural resources with vague property rights, which are difficult to define, and huge externalities, they should continue to be the owners of public property rights. Secondly, improve relevant legislation to provide legal guarantee for the marketization of natural resource property right reform, with emphasis on the subject of resource asset property right transaction, property right transaction rules and property right auction. Finally, by establishing the ownership system of natural resources, the development and utilization of natural resources in China will get rid of the shadow of ineffective and inefficient systems and achieve remarkable system performance. This is the inevitable choice and result of China's market economy. It is also the inevitable choice and result of natural resources from high-cost secret transactions to low-cost public transactions.

The market is an invisible hand to allocate resources effectively. The biggest advantage of the market is efficiency. The market is the direction of China's resource property right reform. Although the publicity and externality of environmental resources and the market mechanism itself, the market mechanism cannot effectively allocate environmental resources and there is "market failure," but it cannot deny the market orientation of China's resource property right reform. While reforming the market, it should strengthen management, strengthen the national macro-control function, reduce "market failure," and thus establish a sound environmental resource property right system.

7.2.2.1   Incentive System

In the *Law of Sand Prevention* provisions of Article 8: "Units and individuals that have made outstanding achievements in combating desertification shall be commended and rewarded by the people's government; Those who have made outstanding contributions to the protection and improvement of ecological quality shall be given heavy awards." These provisions require the State Council, government departments, and relevant departments to reach a consensus, formulate incentives and recognition measures, and establish relatively stable incentive funds to play the role of encouraging society to participate in desertification prevention and control, to play the subjective initiative of all aspects, and to mobilize the enthusiasm of relevant personnel. Incentives include not only financial and policy support but also spiritual encouragement.

In terms of preferential policies, land policies should be further liberalized. Desertified land shall be subject to the contract responsibility system. Ownership shall belong to the collective, and the right to use shall belong to the individual. The policy of "who develops, who uses, who invests, and who benefits" has been implemented. Farmers and herdsmen are drawn from mountainous areas (grassland and desertified land) or responsible mountainous areas (grassland and desertified land) with limited treatment. Allow inheritance and transfer of governance achievements according to law. Some desertified land can be auctioned, leased, and transferred. Encourage domestic and foreign entities to contract and control desertification land.

In terms of financial support, desertification control, large-scale development investment, engineering construction investment, and loan subsidy, the central government should increase financial transfer payments in desertification areas. For important ecological areas, the funds needed for the national antidesertification project approved and implemented by the State Council mainly come from the central finance. The central agricultural, forestry, water conservancy, animal husbandry, energy and other industrial departments, poverty alleviation, comprehensive agricultural development, and other departments should raise funds for unified use and increase the intensity and scale of governance and development. For the protection of desert vegetation, the planting of firewood in the sand and the development and utilization of solar energy, wind energy, biogas, and other energy sources, the state shall give appropriate subsidies. We will continue to implement a stable financial discount policy for sand prevention and control credit and increase the amount of discount loans for sand prevention and control. For fast-growing and high-yield forests, economic forests, ecological agriculture, water-saving irrigation, and small-scale development projects of farmers and herdsmen in sandy areas, long-term preferential interest rate credit support will be implemented to extend the loan discount period. Simplify loan procedures, reform existing mortgage methods, and relax loan conditions.

Under the tax exemption system, units and individuals invest in afforestation, grass planting, and other desertification control activities. It is suggested to exempt all relevant taxes within 3 years after receiving the income. For production enterprises that have obvious effects on desertification prevention and can directly promote vegetation construction and protection, they will be exempted from enterprise income tax for 3 years. Income from engaging in desertification control development, technology transfer, and technical consulting services shall be exempted from business tax and income tax. Nonprofit units such as desertification control experimental bases, experimental sites, and demonstration areas shall be exempted from business tax upon the approval of the people's governments at or above the county level. For the introduction of foreign donations of personal instruments and equipment for combating desertification, exempt from customs duties and import value-added tax, etc. (Fig. 7.1).

7.3 Enterprise Participation

7.3.1 Land Degradation Enterprises

Under the background of market economy, the consciousness of existence and development makes people wake up and begin to seek ways and means to combine ecological construction with their own existence and development. In the prevention and control of land degradation, enterprises have realized a major

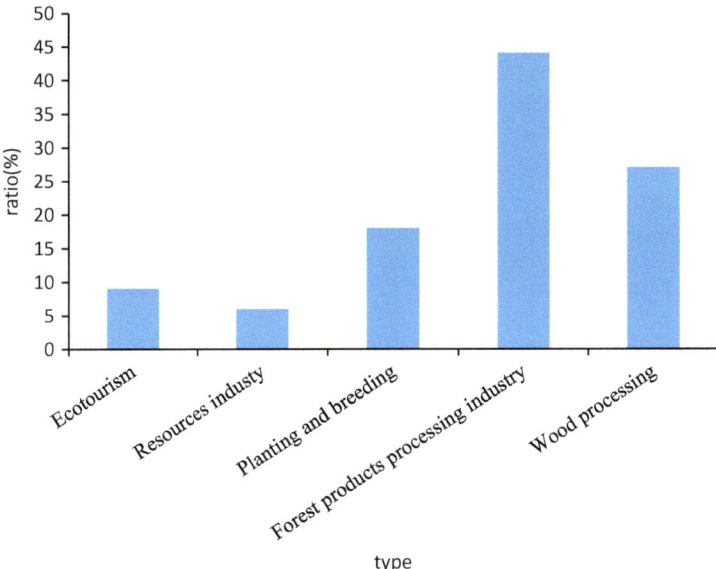

Fig. 7.1. Region to participate in land degradation prevention and control enterprise statistics

transformation from pure ecological type to economic type and found a win-win way of "government tax, enterprise efficiency, and increasing the income of farmers and herdsmen." A large number of leading forestry enterprises have emerged, which have promoted the structural adjustment of forestry and animal husbandry in sandy areas, increased the income of farmers and herdsmen, and injected vitality into the prevention and control of land degradation.

In the public-private partnership, both parties have the role of cooperation and need to give full play to their respective advantages. Public institutions and the private sector have their unique advantages. The adoption of public-private partnerships aims to enable the two partners to develop their respective advantages and maximize their respective advantages.

Judging from this advantage, the public sector can formulate appropriate policies to provide policy support according to needs. Have a good support ability. However, the shortcomings of the public sector are insufficient funds, backward management, and low efficiency. The main advantages of the private sector are relatively abundant capital, advanced management mode, rich management experience, strong flexibility, and strong innovation ability, but the ability of private institutions to bear risks is limited. Because of their different advantages and disadvantages, they have formed different combinations and specific forms of cooperation.

The distribution of resources and site conditions in this region is different. A large number of enterprises are involved in the prevention and control of land degradation. As a result, planting, aquaculture, forestry and fruit industry, purchasing, tourism, and forest products processing industries have developed rapidly in this region. Some have already formed industrialization and become pillar industries of local economic development. According to investigation and statistics, there are 76 enterprises directly involved in the prevention and control of land degradation in the region, including 19 timber processing industries, 7 economic tourism industries, 14 species breeding industries, 32 forest by-product processing industries, and 4 resource industries, with a total of 4.4917 million mu of affiliated interest bases, of which 2.118 million mu (42.79%) is the agreed operating base area.

7.3.2 Public-Private Partnership-Driven Mechanism

Based on relevant theories such as economics, geography, management, psychology, etc., this paper puts forward empirical assumptions on the driving factors of corporate environmental governance behavior and the theoretical framework of public-private behavior driven by public-private partnership.

Location Condition In the process of production and operation activities after the location of the enterprise, the location conditions also have certain influence and restriction. Enterprises are a social group and cannot exist independently. The enterprise is in a certain area. It is necessary to contact and communicate with natural materials and social organizations in the region. Accepting the natural and social environment in the region will cause environmental pressure, which may not be direct, but indirect and potential, but also cannot be ignored.

The influence of location conditions on the environmental behavior of enterprises can be analyzed from two aspects of natural environment and social environment, both of which are caused by different locations. In the natural environment, natural resources are scarce, the ecological environment is fragile, and the development of enterprises is restricted by resources. As raw materials and energy are converted into higher wood, enterprises often pay attention to their own environmental impact and improve their environmental behavior. In the case of abundant natural resources, it is more convenient for enterprises to obtain alternative resources, which are conducive to environmental protection. In the social environment, enterprises are located in areas with high awareness of social environment and have higher requirements on community environment. Enterprises are under greater environmental pressure and are more likely to take environmental actions and vice versa. Enterprises often ignore their own environmental impacts. In addition, the process of enterprises taking environmental actions is to consider environmental protection funds, level, traffic conditions, cultural and educational level, information support, financial service management, and the availability of scientific and technological talents. The higher the level of social and economic development and education, the more enterprises tend to pay attention to environmental behavior. The more information, financial and legal services support, and management and technical personnel are available, the better the transportation network and other infrastructure, and the higher the possibility of contact with other enterprises and other social organizations, the more resources enterprises can obtain to help better adopt environmental behavior.

Under different geographical conditions of natural conditions and social and cultural environment, the eastern part of the country is flat and rich in climate, water, oil, iron ore, and other resources and has good conditions for agricultural development. In addition, this area has a long history of development, advantageous geographical location, developed transportation network, high cultural quality of workers, high level of technical management, and strong industrial and agricultural foundation, plays a leading role in China's economic development, and has the best overall location conditions. The western region is vast in area and rich in mineral resources, but the terrain is relatively high, the terrain is complex, the desert and grassland are white, the construction of transportation network is very difficult, the development of resources is difficult, most regions are cold and lack of water, the development of agriculture is greatly restricted, and due to the late development, the level of economic development and technical management is low, the overall location conditions are poor. On this basis, the following assumptions are put forward:

Hypothesis 1 Enterprises in the eastern region have better macro-location conditions, and their public and private behaviors are more developed than those in the central and western regions.

Environmental Regulation Pressure Supervision pressure mainly refers to the government's intervention and intervention on the environmental behavior of enterprises based on the formulation of laws and regulations. Environmental regulation is directive and mandatory and is the bottom line of enterprise behavior. If an enterprise violates relevant laws and regulations, it will lose the legitimacy of existence. Rules and regulations are reward and punishment measures to guide enterprises to attach importance to environmental protection and improve environmental behavior. In order to reduce illegal punishment on wood, enterprises must abide by rules and regulations to ensure good environmental behavior relationship with the government. In addition, environmental regulation can also guide the public's consumption patterns and habits, adopt market means to promote enterprises to produce green products, and improve environmental behavior, while the operation and implementation of environmental regulation have an impact on the compensation mechanism.

For the social groups of enterprises, the government and society will intervene from the outside, of which government supervision is mandatory and public opinion has greater influence on enterprises. Government intervention in enterprises' environmental behavior is an activity taken by the government in order to improve the environmental performance of enterprises and the whole society, for example, taking policy guidance, system supervision, technical funds, and incentive measures.

Supervision is divided into command and control supervision and incentive supervision.

We can think that government regulation is an important driving force for enterprises' environmental behavior, and different types of regulation have different driving effects on enterprises to improve their environmental behavior. Incentive supervision is more conducive to improve the efficiency of enterprises, which can adopt the trend of coping strategies, improve the internal goal setting of enterprises, and further improve the environmental behavior level of enterprises.

Enterprise Motivation The internal motive of an enterprise is the motive of the enterprise. There are two kinds of motives. The most definite one is to maximize profits. This is driven by the original motive of the enterprise's continuous development and growth. However, with the continuous increase of the current social demand, if an enterprise avoids fulfilling its social responsibility, it cannot survive in the market competition, which will lead to the loss of its social rights. Based on the above theory, we can think that enterprises will make environmental behavior choices based on different motives, such as the pursuit of profit maximization and the need to assume social responsibility. The growing green consumption market and the green industry with broad development prospects show that enterprises can obtain good market opportunities and corresponding profits. By developing appropriate environmental protection projects and adopting technological innovation and resource recovery measures, enterprises implement reasonable environmental protection behaviors, reduce the use of resources and wastes, realize cost savings, improve product quality and functional price, create economic benefits, and enable enterprises to occupy a dominant position in the competition. However, in the face of increasingly serious environmental problems, enterprises have a strong sense of social responsibility, with their own survival and long-term development as the ultimate goal, and consciously undertake social responsibilities, including environmental governance and balance, and coordinate the relationship between their own interests and environmental public welfare and also establish a good corporate image and obtain a comparative competitive advantage. Obviously, profit motive and environmental responsibility are the main driving forces for enterprises to actively respond to government environmental policies and adopt environmental behaviors. Based on this, the following two assumptions can be made:

Hypothesis 2 The stronger the enterprise pursuing economic benefits, the greater the possibility of taking environmental protection actions and the better.

Hypothesis 3 The stronger the motivation of corporate social responsibility, the greater the possibility of taking environmental action and the better the environmental behavior.

Enterprise Industry Characteristics Today's enterprises often go hand in hand in various industries, and it is difficult to develop on the basis of industrial management types for a long time. Especially in my region, there are many resource-based enterprises due to abundant plant and mineral resources. In other words, the enterprise itself acts to prevent land degradation, so we can make this assumption.

Hypothesis 4 As long as enterprises participate in the resource industry, the more likely they are to adopt environmental behaviors, the better the environmental behaviors will be.

7.3.3 Relationship Status of Public-Private Partnership

In fact, degraded land is not only a carrier of governance but also an available resource. In order to make the most reasonable allocation of barren hills and wasteland, the management should give full play to the potential resources of barren hills and wasteland in the process of land degradation, adjust measures to local conditions, make rational use, and make the best use of land resources. In China's rural areas, there are objective differences in production conditions and imbalance in industrial development. The protection and prevention of regional land resources should not be forced to unify the fixed mode. In order to adapt to the different ways of land resources in different regions and farmers in the process of land degradation management and industrial structure adjustment, there are many flexible land resources management modes in different small regions.

7.3.3.1 Questionnaire Design

On the basis of previous literature research, with the final purpose of investigating the current situation and behavior-driven mechanism of public-private partnerships, the research designed a questionnaire (Appendix) asking about the basic information of enterprises, the implementation of land degradation control measures, the internal and external environment of public-private partnerships to control land degradation, and its impact.

The issues of interest in this article have been basically covered. In the questionnaire design, as much as possible, the common measurement problems in the existing literature are used.

Part 1 Basic information about the company. Including name, number of employees, ownership and type of functional area in which the enterprise is located. The design of these projects conforms to the enterprise characteristics in the internal dynamic factors and the enterprise location conditions in the external pressure factors. The characteristics of a company are measured by the number of employees and the type of ownership. The main measure of enterprise scale is the number of employees, total assets or sales, plus the number of employees and total assets, and the enterprise scale is counted. The name of the enterprise and the type of the main body of the enterprise provide the macro and micro location information of the enterprise, respectively, which can analyze the location conditions of the enterprise according to the actual situation of our country.

Part 2 Environmental awareness of enterprise managers. This paper discusses the environmental awareness of enterprise managers from the aspects of environmental ideology, environmental-related knowledge, and behavior.

Part 3 On the relationship status of public-private partnership. Investigates environmental investment behavior and comprehensively grasps the current situation of purchasing power parity behavior. To investigate the environmental planning policies of enterprises, the establishment of environmental objectives, the setting up of environmental protection agencies, the development of environmental investment, environmental education and training, and to participate in social environmental protection activities.

Influencing factors of public-private partnership. Motivation, internal environment, and external environment come from the formation of public-private partnership, including cost motivation, income motivation, motivation to enhance competitiveness, motivation to improve the relationship with the government and other stakeholders, motivation to avoid risks, motivation to break the deal, and motivation to assume social responsibility. Environmental regulation in external pressure factors of enterprise environmental behavior includes perfection of national environmental protection regulation, importance of local government to environmental protection, severity of local law enforcement, preferential policies for local environmental protection, and changes and requirements of environmental regulation.

Each question in the questionnaire and its various investigation items adopt Likert's five-point method. According to the situation of the problem, the options of "totally disagree" to "totally agree" and "totally disagree" to "totally think" are given. Ask respondents to check the corresponding options and assign values to each option in the subsequent data processing stage.

7.3.3.2 The Results of the Survey

In early July 2011, the author sent 815 questionnaires to 144 selected companies. As of early February 2012, 601 questionnaires had been collected. All questionnaires were valid, with a recovery rate of 73.7% and an effective rate of 100%.

After the questionnaire is withdrawn, from "totally disagree" to "totally agree" and from "totally disagree" to "totally agree" from low to high, giving 1–5 points, carefully input data and check one by one, excluding unnecessary questionnaires, incomplete data, false filling, recording questionnaire data, and basic information of statistical sample information (see Table 7.1).

According to this survey, the industries of the sample enterprises are resource industry, ecotourism, forestry, wood product processing, and aquaculture, among which there are many kinds of aquaculture, and the industrial enterprises account

Table 7.1 Sample information data statistics

Project	Classification	Number	Percentage (%)
Industry type	Forest product	34	23.61
	Resources industry	28	19.44
	Resources industry	12	8.33
	Wood processing	24	16.67
	Aquaculture	46	31.94
Enterprise scale	Large-sized	14	9.7
	Medium-sized	86	59.7
	Small-scale	44	30.4
Ownership properties	State-owned and collective	0	0.00
	Joint venture and foreign capital	0	0.00
	Privately operated	144	100.00
Enterprises belonging to the region	East	58	40.00
	Middle	37	26.00
	West	49	34.00

for 31.94% of the total sample. According to the enterprise scale, there are 86 largest medium-sized enterprises, accounting for 59.7% of the total sample, while only 14 large enterprises, accounting for 9.7% of the total sample. According to the nature of ownership, they are all private enterprises, accounting for 100% of the total sample. According to the macro position of enterprises, there are 58 enterprises in eastern China, accounting for 40% of the total, 37 enterprises in central China, accounting for 26%, and 49 enterprises in western China, accounting for 34%.

7.3.3.3   Current Situation Assessment

Based on the data from the survey on the status quo of public-private partnership behaviors in the questionnaire, this study selects seven behavioral status evaluation indicators D1–D7, namely, the status of environmental target setting, the status of land degradation prevention and control departments, the status of environmental performance, the status of environmental investment, the status of environmental education and training, the status of environmental protection activities, and the status of implementation of environmental information disclosure. This paper uses Cronbach test to test the reliability of the scale. Cronbach coefficient is internal consistency, and its value is between 0 and 1. Alpha >0.7 indicates that the reliability of the scale is high and the questionnaire can be accepted. If 0.7 >alpha >0.35, the reliability of the scale is good and is also acceptable. Alpha <0.35 indicates a low reliability and is unacceptable.

Reliability tests were carried out on seven evaluation indexes and observation items of this study, and the results were obtained. As shown in the table, the Cronbach alpha for each observation item is 0.85 which is greater than 0.7, indicating that we have chosen seven observation items with high internal consistency and reliability (Table 7.2).

Table 7.2 Sample reliability monitoring

Indicator code	Evaluating indicator	Evaluating indicator	Average value	Standard deviation	Cronbach (α)
Gross list					α = 0.85
D1	Status of environmental performance	100	4.17	0.877	
D2	Status of environmental input	98	4.01	0.959	
D3	Environmental goal setting	105	4.43	0.936	
D4	Status of land degradation prevention and control department	110	4.85	0.849	
D5	Environmental education and training	101	3.52	1.003	
D6	Status of environmental protection activities	101	4.03	0.913	
D7	Status of environmental information disclosure	101	2.63	1.316	

The correlation coefficient and its significance are used to reflect the correlation between the seven environmental behavior evaluation indexes. The correlation coefficient indicates the degree of correlation between each evaluation index and other indexes. The smaller the correlation coefficient is, the smaller the correlation between this index and other indexes is. The higher the significance, the more significant the correlation between the index and other indicators. According to the practical significance of each evaluation index, the evaluation with small correlation coefficient and significantly low correlation coefficient is eliminated.

The table shows the correlation coefficient between each evaluation index of enterprise environmental behavior. It can be seen that the correlation coefficient between each index is large and significant, while the correlation between D7 and other factors is small and significant. Among them, D7 represents that the status of "environmental information disclosure" in the behavior is not very high, so we removed this evaluation index (Table 7.3).

Table 7.3 Correlation coefficient of evaluation index

	D1	D2	D3	D4	D5	D6	D7
D1	1						
D2	0.547*	1					
D3	0.395**	0.665**	1				
D4	0.401**	0.673**	0.373**	1			
D5	0.231*	0.216*	0.343**	0.288**	1		
D6	0.524**	0.349**	0.284**	0.447**	0.366**	1	
D7	0.066	0.171	0.221*	0.285**	0.280*	0.174	1

**It shows that the correlation coefficient of variables is significant at 0.01 level, and the correlation coefficient of variables is significant at 0.05 level (double-tailed test)

Table 7.4 Principal component analysis results of each index

Principal component	Initial eigenvalue			Rotating square and loading		
	Total	Variance explained (%)	Cumulative variance explained (%)	Total	Variance explained (%)	Cumulative variance explained (%)
1	3.406	64.231	64.231	6.987	49.487	49.487
2	1.345	12.657	76.888	4.654	27.401	76.888
3	0.887	9.897	86.785			
4	0.805	5.302	92.087			
5	0.585	4.235	96.322			
6	0.549	3.678	100			

Factor analysis in this study mainly adopts principal component analysis extraction method and Kaiser standard orthogonal rotation method. The results are as follows (Table 7.4).

Using Kaiser criterion, principal component factors with eigenvalues greater than 1 are extracted, and two principal component factors with eigenvalues greater than 1 are extracted. The cumulative variance interpretation rate of these two principal components reached 76.888%, which explained the structure of the questionnaire. The effectiveness meets the requirements (generally 55%) (Table 7.5).

The results show that enterprise location, environmental regulation pressure, enterprise scale, and ownership characteristics have significant driving effects on enterprise environmental behavior, and relevant assumptions are supported, while enterprise macro location, stakeholder pressure, leadership environmental awareness, and economic interest pursuit have significant driving effects on enterprise environmental behavior. The hypothesis that motivation and social motivation drive the environmental behavior of enterprises is not supported.

7.3.4 Reasons for Enterprises to Participate in Land Degradation Control

In order to obtain the overall characteristics, site conditions, implementation, and results of the case investigation points in this area, a questionnaire was prepared through the collection of technical, socioeconomic, financial, and related data. Use interviews with senior leaders, consulting project reports, and other available literature methods to obtain the information required by the questionnaire. Through the analysis of the factors affecting the internal and different projects and the project results, the following four main reasons are obtained:

- The initial driving force of some private sectors is to obtain stable and low-cost raw materials.
- Larger integrated companies are driven by deepening resources, reserves, and capital appreciation.

Table 7.5 Descriptive statistics and reliability test of behavior-driven factor questionnaire

	Sample size	Sample size	Sample size	Minimum value	Maximum value	Cronbach α
Leadership environmental awareness						
1. Environmental protection is the obligation of citizens	101	4.76	0.650	1	5	α = 0.902
2. Environmental protection is the obligation of citizens	101	4.61	0.632	1	5	
3. Pay attention to all kinds of environmental problems	101	4.39	0.707	1	5	
4. Understanding of China's current environmental policies and regulations	101	4.93	0.840	1	5	
5. Pay attention to energy saving and emission reduction in life	101	4.29	0.753	1	5	
6. Implementation of energy-saving emission reduction work	101	4.38	0.746	1	5	
Environmental regulation pressure						
1. National environmental protection rules	99	3.94	0.867	2	5	α = 0.830
2. Place emphasis on environmental protection	99	4.04	0.820	2	5	
3. Local law enforcement and justice	98	3.84	0.884	1	5	
4. Place the introduction of preferential policies for environmental protection	100	3.76	0.874	2	5	
5. Nearly 5 years of environmental rules change	99	3.86	0.769	2	5	
6. The change of environmental rules can be predicted	99	3.60	0.755	2	5	
7. Environmental regulations on enterprise requirements	99	3.83	0.770	2	5	
Stakeholder pressure						
1. Environmental protection requirements of corporate shareholders	99	3.77	0.767	2	5	α = 0.734
2. Customers, consumer demand for environmental protection	98	3.89	0.758	2	5	
3. Competitors are in the implementation of energy-saving emission reduction	99	3.73	0.843	2	5	

(continued)

Table 7.5 (continued)

	Sample size	Sample size	Sample size	Minimum value	Maximum value	Cronbach α
4. Surrounding community environmental protection requirements	100	4.03	0.674	2	5	
Motivation to pursue economic interests						
1. Reduce the cost of fines and sewage charges lower production cost	101	2.95	1.186	1	5	α = 0.837
2. Obtain economic benefits	101	3.83	1.068	1	5	
3. Enhance the competitiveness of enterprises	101	3.85	1.014	1	5	
4. Improve the relationship with the government	101	4.42	0.667	2	5	
5. Meet the requirements of stakeholders	101	3.52	0.965	1	5	
6. Avoiding environmental risks	101	3.66	1.023	1	5	
7. Breaking the green trade barrier	101	4.03	0.921	2	5	
	101	3.64	0.996	1	5	
The motive of taking social responsibility	101	4.56	0.590	2	5	

- There are also large entrepreneurs (foundations, etc.). The goal of this project is to establish an image through the dedication of public welfare undertakings, the restoration of ecological environment, and the recognition of the government and the people.
- Preventing land degradation is not only a government-led public welfare activity but also an enterprise-centered economic activity.

Chapter 8
Public-Private Partnership Practice Case Analysis

8.1 The Case of Yili Group

Yili Resources Group's original production base is located in the abdomen of Kubuqi Desert, which has suffered greatly. Specifically, the management of the Kubuqi Desert was forced to withdraw. At that time, because Haizi, the birthplace of Yili Resources, was blocked by Kubuqi Desert, the transportation of products and raw materials had to detour 330 kilometers to reach the railway station. Therefore, the annual cost will exceed 15 million yuan, almost offsetting the company's profits. The Kubuqi Desert makes it difficult for local farmers and herdsmen to see a doctor and go to school and shop. It has become the biggest obstacle for people in the desert to become rich. In order to protect lake mineral resources such as trona, thenardite, and salt, which the group depends on for its survival, Yili Resources set up a forestry fund as early as 1988 to extract 15 yuan per ton of products for ecological construction and set up a forestry team. A forestry team specialized in research, afforestation, and management has been established. At first, only sand barriers, Caragana, willow, and poplar were set up by the lake. From 1997 to 1999, Yili Resources invested in the construction of the first sand-crossing highway. Due to the fluidity of the Kubuji Desert, it is impossible to stop sand or protect roads. In this way, Yili Resources were forced to take the road of preventing and controlling desertification, which opened the prelude of Yili Resources desert control and sand industry development.

In the past 20 years, Yili Resources Group has implemented six major desertification control projects:

First, five roads with a total length of 234 kilometers were built, and sand control methods of "roads, electricity, water, communication, network, and green" were explored.

Second, a sand control and shelterbelt project with a length of more than 242 kilometers and a width of 3–5 km has been implemented in the narrow strip along the northern edge of Kubuqi Desert and the southern bank of Yellow River.

© Science Press & Springer Nature Singapore Pte Ltd. 2020 217
Z. Meng et al., *Public Private Partnership for Desertification Control in Inner Mongolia*, https://doi.org/10.1007/978-981-13-7499-9_8

Third, vigorously develop the traditional Chinese medicine processing industry based on desert licorice, reverse the situation of desertification prevention and control, and promote the common people to shake off poverty and become rich.

Fourth, use the natural scenery of the desert and ecological construction results, combined with grassland, desert natural landscape, and Chinese herbal medicine base to develop ecological tourism, and build the Kubuqi Desert Qixing Lake tourist area.

The fifth is to carry out ecological immigration to the herdsmen scattered in millions of mu of sandy land so that they can live in a "new village for herdsmen" where intensive production, living, management, and development are carried out. At the same time, 1500 square kilometers of "desert ecological fences and protected areas" will be built.

The group has a typical public-private partnership in the following respects:

8.1.1 Desert Tourism Industry

This project is the golden signboard industry of the group based on desert management and developing desert economy. The project is based on public asset desert land. Through government loans, enterprise design, and management, it will drive people around to become rich and improve the local ecological environment. A typical public-private partnership structure is as follows (Fig. 8.1):

In this typical public-private partnership model, the project agreement defines the rights and responsibilities of the public sector and the group in providing public services, as well as the service level and the cost of the payment mechanism. The construction agreement is usually a fixed-price turn-key contract. Construction risks are mainly determined by contractors. Operation and maintenance are usually divided into some professional operation companies and property maintenance companies. Most of these private sectors bear the operation and maintenance risks (see figure).

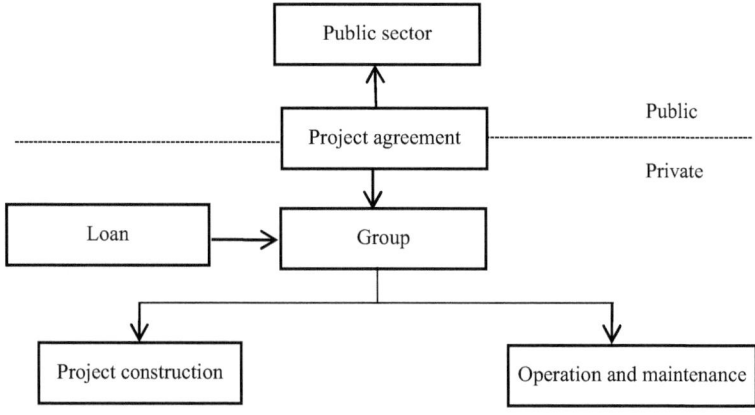

Fig. 8.1 Structure of typical PPP model

8.1.2   Ecological Migration Project

In order to make rational use of the renovation of their homes and enable the local people to lead a happy life, Yili Group has implemented a popular project-ecological immigration, the establishment of Dugupta New District. In this process, a total of three participants participated in the establishment of Brilliance Project, namely, group companies, farmers, and the government. In short, in this project, the company signed a cooperation agreement with the farmers, clarifying the responsibilities, rights, and interests of both parties on the basis of equality and mutual benefit. During the cooperation period, the company and the farmers jointly bear risks and benefits. The government and the company set up a management committee to coordinate the project and ensure its smooth completion. Participants played different roles in the project. The company is the core of the model and is responsible for organizing production, making decisions, and providing production facilities. Before production, the company should provide farmers with high-quality means of production and infrastructure. The main task of farmers is to use the company's production facilities and their own means of production (such as land, etc.) to carry out production activities. In some cases, farmers can also exchange their land for shares, capital, or reward by providing other means of production. The government is the coordinator between the company and the farmers. On the one hand, the government provides various preferential policies for companies to reduce policy risks and ensure farmers' production, thus attracting investment companies. On the other hand, the government should also negotiate with investors to supervise the progress of the project, mainly responsible for project management, including promoting communication, strengthening organization and supervision, and providing technical support. The three participants are linked through public-private partnerships to jointly organize local resources and local economic development (see Fig. 8.2).

8.1.3   New Oriental School Education

In July 2009, the group donated 1.1 billion yuan for the construction of "Yili Oriental School" in Hangjinqi. The school adopts a private management system. The team is responsible for the daily management and operation of the school. The Local Education Bureau is responsible for supervising and recruiting school graduates (Table 8.1).

8.1.4   Yili Infrastructure Construction Project of the Yellow River Bridge

In July 2010, an investment of 300 million yuan was invested to build the Yili Yellow River Bridge in the desert. The public-private partnership project is the market operation of the project and requires a reasonable return on investment. At present, road and bridge tolls are subject to a unified government pricing mechanism.

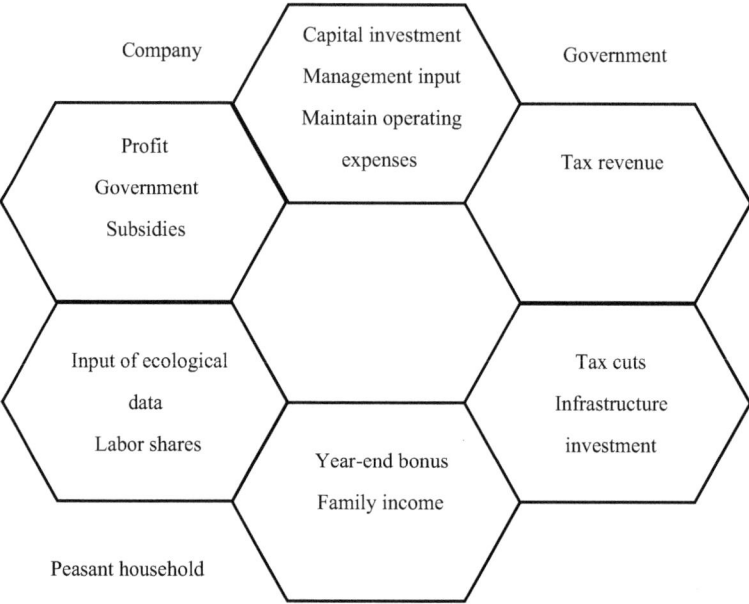

Fig. 8.2 Public-private partnership model

Table 8.1 Examples and basic features of public-private partnership action plan for basic education

School private action plan	Charter schools	Charter schools operate less than public schools but must meet more performance requirements
		Free school for students
		The school is run by the group

The Status of the Government in the Public and Private Cooperation of Road and Bridge Project The government's main responsibility is to play a leading role in policy formulation, increasing investment and strict supervision to ensure social equity in health services, strengthening health supervision and law enforcement functions, boosting anti-monopoly and anti-dumping regulations, monitoring of service prices, prevention of price fraud and other acts of unfair competition, and protection of the people's legitimate right to health in accordance with the law.

The company's position in public-private cooperation of road and bridge projects:

1. To assume the role of the investor
2. To assume the role of financing
3. Direct interest income

Status of Farmers the ultimate benefit of the project target groups and the direct beneficiaries of the project are also the consumer group.

8.2 Comparison of Project Financing Mode

The project is appropriately separated according to the principle of "investment, construction, management, and use." The source of funds is government financial allocation and commercial loans, and the source of repayment of principal and interest of loans is income from other industries of the group. This project adopts a typical public-private financing mode, with Yili Resources Group Co., Ltd. as the main investor. The government has issued preferential policies for land resources.

8.2.1 Project Organization Operation Mode

The preliminary work of the project is initiated by the local government, the local development and reform commission, the finance bureau, the planning bureau, and the departments of land, environmental protection, and fire protection. With the strong support of the previous government, the enterprise is responsible for the completion of the later construction work. The enterprise is the main body of investment, which is a typical "investment-construction-operation" mode and design. After the construction work is completed, the enterprise is responsible for the project operation, performs the responsibilities of the owner, manages the project contract and information, organizes and coordinates the project according to the plan, and accepts the supervision of relevant departments dispatched by the government. The specific project organization structure can be summarized as follows (Fig. 8.3).

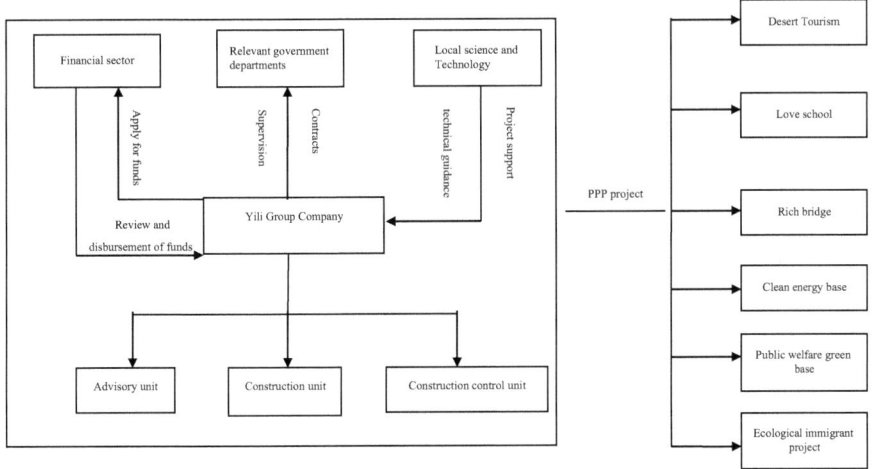

Fig. 8.3 Yili Group project organization structure

8.3   The Key Factors of Yili Group PPP

The focus group first adopted in-depth interviews. Select some people from the top management, office, grass roots, and pastoral areas to understand their views on the development of public-private partnership and find out the main factors affecting the development of public-private partnership.

The research strategy of any research is to a large extent the research purpose, the nature of the research problem, the available resources, and the technology used. Since the public does not have PPP-related knowledge, the interviewee can only be part of the crowd experience. Because of interviewing specific people, the content of the interview is very professional, there are many problems, and it takes a long time. Taking this form of interview, the interview time is controlled within 0.5–1 h according to the specific situation. The research includes three questions in the interview: Which factors are conducive to the development of purchasing power parity? What are the obstacles to the development of public-private partnerships? What are the conditions for a successful public-private partnership? The interview was conducted by random sampling, and the sample distribution is shown in the following table (Tables 8.2 and 8.3).

In the public sector, local government officials, forestry institutions, and township staff are mainly selected, while the private sector mainly refers to senior management of enterprises and frontline staff at the grassroots level. After sorting out and analyzing the interview content, as long as there is one factor related to the respondent's vocabulary in the above three questions, it is considered to be a definite factor, only mentioned once, and finally the number of each factor is calculated. The overall structure of the interview is shown in the following table.

By comparing the results of these interviews, we can find that the perfection of legal system and policies, identity value, and government support ranks first. "Perfect laws and regulations" refer to the laws that have been determined, are consistent with laws and regulations, can be specifically enacted for public-private part-

Table 8.2 A list of the number of interview samples

Public sector	Business sector	Nonprofit and non-governmental organizations
12	10	4
8	6	2
20	16	6

Table 8.3 Key conditions for developing public-private partnerships

Factor	Frequency	Factor	Frequency
Political stability	3	Government approval process is simple	1
Perfect law and system	21	Reasonable risk sharing	9
Identification of values	16	High degree of economic development	3
Government support	6	Can obtain capital	11

nerships, and can also be included in other laws. "Identity value" can reflect the wishes of political leaders as well as people's ideology. "Government support" mainly refers to the willingness and ability to abide by contractual commitments and the mechanisms and measures to ensure the implementation of government credit and public-private partnership projects. "Capital is available." In fact, this reflects that the special requirements of the national capital market are not perfect. Only when local capital is sufficient can the financing cost of public-private partnership projects be reduced. "Risk sharing" is an important aspect of reducing uncertainty and risks in the private sector (Fig. 8.4).

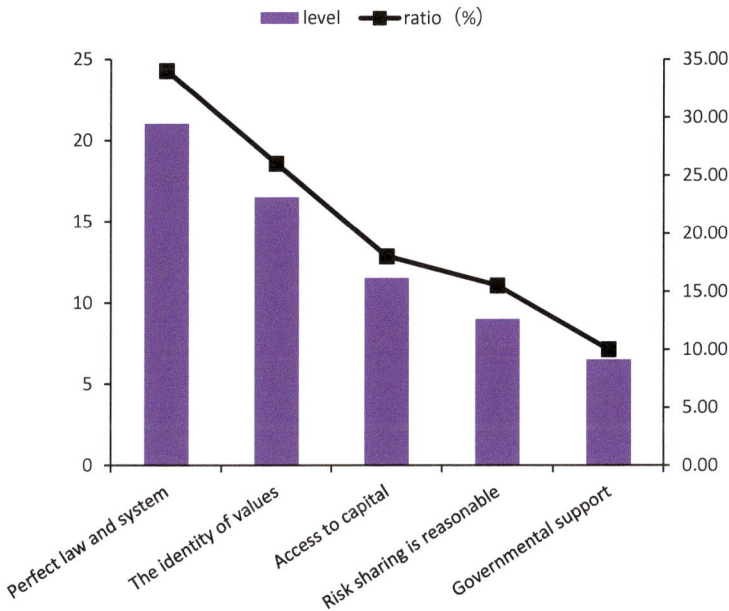

Fig. 8.4 Key factors affecting the success of public-private partnerships

Chapter 9
Conclusions and Suggestions

9.1 Conclusions

The prevention and control of land degradation has strong public welfare and externality, and the possibility of full marketization is very small. The implementation of public-private cooperation mode is an ideal way to guide enterprises and individuals to participate in the ecological environment. The introduction of public-private cooperation in the prevention and control of land degradation in our district can achieve a win-win situation (Fig. 9.1).

As far as enterprises are concerned, expanding new investment channels can bring greater benefits, mainly economic benefits, in the long run. For farmers, through the contribution of labor, improve their production and living conditions and further improve the lives of herdsmen. From the perspective of the government, it can significantly reduce government investment and improve investment efficiency.

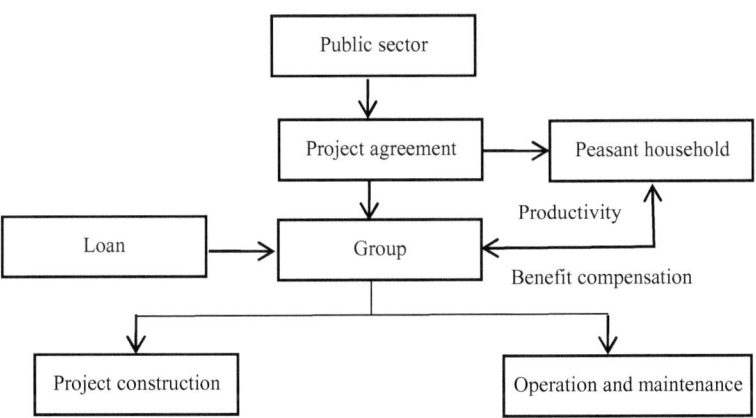

Fig. 9.1 Block diagram of public-private partnership in our district

© Science Press & Springer Nature Singapore Pte Ltd. 2020
Z. Meng et al., *Public Private Partnership for Desertification Control in Inner Mongolia*, https://doi.org/10.1007/978-981-13-7499-9_9

Judging from the application of purchasing power parity model of land degradation in our region, the implementation of the project is easier. There are many differences between regions in the eastern developed regions. At present, the public-private cooperation in our district has mainly formed an eco-economic model of "government+company+base+farmers and herdsmen."

9.2 Suggestions

PPP is more and more widely used. China has great potential to develop public-private partnerships in many industries and fields. However, there are also many problems and shortcomings. In order to further promote the development of public-private partnerships and expand the participation of land degradation prevention teams, the following recommendations are made: improve the public participation mechanism of China's PPP project and improve the participation of stakeholders.

We will improve the financing environment, improve relevant laws on project financing, and establish a set of laws and regulations on project financing that are in line with China's national conditions.

Appendices

Appendix A

Investigation on the status quo and related conditions of public-private cooperation in Inner Mongolia Autonomous Region

You fill in relevant information in this line and check the options.

The name of firm		Number of enterprise employees			
Enterprise ownership	☐ Enterprise main body is located	☐ The collective	☐ Foreign	☐ Private	☐ Other
Enterprise main body is located	☐ The economic development zone	☐ Industrial zone	☐ Business district	☐ Mixing zone	☐ Other

Personal Survey

Do you agree with the following statement regarding your personal attitude?				
Environmental governance is every citizen's obligation	☐ Totally disagree	☐ Disagree	☐ Neutral	☐ Agree
People are very concerned about the land degradation treatment	☐ Totally disagree	☐ Disagree	☐ Neutral	☐ Agree
People are very concerned about the problem of desertification	☐ Totally disagree	☐ Disagree	☐ Neutral	☐ Agree
To understand the environmental policies and regulations	☐ Totally disagree	☐ Disagree	☐ Neutral	☐ Agree
Usually pay attention to the protection of the environment	☐ Totally disagree	☐ Disagree	☐ Neutral	☐ Agree
The implementation of green consumption	☐Totally disagree	☐Disagree	☐Neutral	☐Agree

© Science Press & Springer Nature Singapore Pte Ltd. 2020
Z. Meng et al., *Public Private Partnership for Desertification Control in Inner Mongolia*, https://doi.org/10.1007/978-981-13-7499-9

Implementation of Degraded Land Management

Specific environmental governance objectives	☐ No such plan	☐ Under discussion	☐ Being worked out	☐ Just make	
Self-energy consumption and pollution emission	☐ No such plan	☐ Under discussion	☐ Not statistics	☐ Statistics	☐ Statistics, report
Set up special environmental management department	☐ No such plan	☐ Under discussion	☐ Other departments in escrow	☐ Already established	☐ Has been set up, the person in charge
Prevention and control of land degradation is mainly used for	☐ Afforestation	☐ Environmental protection propaganda and training	☐ Fund sponsor	☐ Set up base	☐ Technology research and development
Land degradation control investment accounted for the proportion of total annual income (%)	☐ Below 0.5	☐ 0.5–1	☐ 1–1.5	☐ 1.5–2	☐ More than 2
Participate in environmental activities	☐ No plan	☐ Under discussion	☐ Sometimes involved	☐ Often participate in	☐ Have a plan to participate in

Public-Private Enterprise Motivation

Reduce cost	☐ Totally disagree	☐ Disagree	☐ Neutral	☐ Agree
Economic performance	☐ Totally disagree	☐ Disagree	☐ Neutral	☐ Agree
Enhance competitiveness	☐ Totally disagree	☐ Disagree	☐ Neutral	☐ Agree
Improvement and government relations	☐ Totally disagree	☐ Disagree	☐ Neutral	☐ Agree
Meet stakeholder	☐ Totally disagree	☐ Disagree	☐ Neutral	☐ Agree
Corporate responsibility	☐ Totally disagree	☐ Disagree	☐ Neutral	☐ Agree

Interior Environment

Leadership attention	☐ Totally disagree	☐ Disagree	☐ Neutral	☐ Agree
Employee participation	☐ Totally disagree	☐ Disagree	☐ Neutral	☐ Agree
Reasonable structure of enterprise	☐ Totally disagree	☐ Disagree	☐ Neutral	☐ Agree
Flexible system	☐Totally disagree	☐ Disagree	☐ Neutral	☐ Agree
Strong economic strength	☐ Totally disagree	☐ Disagree	☐ Neutral	☐ Agree
Sound training system	☐ Totally disagree	☐ Disagree	☐ Neutral	☐ Agree
Expert information access channel	☐ Totally disagree	☐ Disagree	☐ Neutral	☐ Agree

External Environment

Perfection of system and regulation	☐ Totally disagree	☐ Disagree	☐ Neutral	☐ Agree
Government attention	☐ Totally disagree	☐ Disagree	☐ Neutral	☐ Agree
Local government introduced preferential policies	☐ Totally disagree	☐ Disagree	☐ Neutral	☐ Agree
Environmental regulations in recent years, a large number of policy changes	☐ Totally disagree	☐ Disagree	☐ neutral	☐ Agree
Regulations on business requirements	☐ Totally disagree	☐ Disagree	☐ Neutral	☐ Agree

Direct Economic Benefit

Reduce the cost of raw materials	☐ Totally think	☐ Not think	☐ Neutral	☐ Think
Improve production efficiency	☐ Totally think	☐ Not think	☐ Neutral	☐ Think
Reduce taxes and interest	☐ Totally think	☐ Not think	☐ Neutral	☐ Think
Direct economic income	☐ Totally think	☐ Not think	☐ Neutral	☐ Think

Indirect Benefit

Enhance image	☐ Totally think	☐ Not think	☐ Neutral	☐ Think
Reduce environmental risks	☐ Totally think	☐ Not think	☐ Neutral	☐ Think
Improve customer loyalty	☐ Totally think	☐ Not think	☐ Neutral	☐ Think

Improve the resident environment	☐ Totally think	☐ Not think	☐ Neutral	☐ Think
Promote the relationship between government and enterprises	☐ Totally think	☐ Not think	☐ Neutral	☐ Think

Appendix B: Interview-Related Issues

1. Do you think public-private cooperation is a way to pay attention to public services and public service provision?
2. If yes or no:

   Question 1: Why?
   Question 2: In which areas of service? Why?
   Question 3: What factors are conducive to the development of PPP?

Interviewer note: Remember some examples to prevent respondents from not understanding or answering questions.
Favorable factors such as:

- Third-party revenue increase
- To obtain financing to enrich the existing funds
- Improvement of service level
- Decreased risk perception

Eager to draw on the technical experience of the private sector
Better use of funds, etc.
What are the obstacles to the development of public-private partnerships?
Interviewer note: Remember some examples, in case the respondents do not understand or answer questions.
Obstacle factors such as:

- The attitude of the senior management or government.
- Lack of documentation
- Lack of training and technical knowledge
- Cost
- Management time
- The complexity of the contract, etc.

3. According to your view, what are the conditions for the success of PPP?

   Question 1: Can you list some of the successful PPP cases in China or other countries?
   Question 2: Can you list some of the unsuccessful PPP cases in China or other countries?

Appendix C: Questionnaire-Related Questions

1. What is your familiarity with PPP?

   (A) Is not familiar with
   (B) Know some
   (C) More familiar
   (D) Is very familiar with

2. How long have you been working in PPP?

   (A) Never
   (B) Below 2 years
   (C) Three to 5 years
   (D) More than 5 years

3. Are you satisfied with the current infrastructure service in China?

   (A) Is not satisfied
   (B) Some satisfaction
   (C) Is more satisfied
   (D) Very satisfied
   (E) Don't know

4. What are the obstacles to the development of public-private partnership projects in your industry? Please use a rating scale of 1 (unimportant) to 4 (very important) to assess the impact of the following barriers on the importance of purchasing power parity (local check box corresponding to "I don't know").

	Not important	Very important	Do not know
(1) Lack of international funds			
(2) Lack of local funds			
(3) The lack of local government guarantees			
(4) Legal norms is weak			
(5) Lack of national level PPP unified management system			
(6) Local government PPP project decision-making process is not clear			
(7) Lack of experience in the development of PPP projects			

5. What measures can be taken to overcome the possibility of public-private partnership development? Please use a scale of 1 (unimportant) to 4 (very important). Assess the importance of the following measures (local check box corresponding to "I don't know").

	Not important	Very important	Do not know
(1) To strengthen the government's control and supervision of the PPP project			
(2) In the development and implementation of PPP project			
(3) Participation in civil society and public organizations			
(4) Organize special seminars, communication research, and practical experience			
(5) Organize seminars, training for civil servants, and businessmen			
(6) The introduction of specialized courses in universities or other educational activities			

6. To what extent do you think the private sector should participate in the public sphere? A should not participate (all infrastructure and public services should be owned and operated by the government).

   (A) The private sector has less than 20% of its operations.
   (B) The private sector has a share of more than 20%, but less than 50%.
   (C) The private sector has a share of more than 50%, but less than 100%.
   (D) Don't know.

7. Do you think the introduction of private capital into traditional public areas will help improve the quality of public services?

   (A) Simply cannot
   (B) Has some help
   (C) More helpful
   (D) Very helpful
   (E) Don't know

8. Does PPP improve the service levels in the following areas? Please use a scale of 1 (strongly disagree) to 4 (strongly agree) to indicate your approval.

	Strongly disagree	Strongly agree	Do not know
(1) Price (price and quality comparison)			
(2) Cost control ability			
(3) Timeliness of service			
(4) Ability to meet expectations of the public			
(5) Infrastructure modernization			

9. Compared with the infrastructure and services provided by the traditional government, do public-private partnerships place more emphasis on protecting the natural environment?

   (A) Yes
   (B) No
   (C) Don't know

10. Compared with traditional government infrastructure and services, do public-private partnerships make more effective use of natural resources (water, oil, natural gas, arable land, forests and timber, etc.)?

   (A) Yes
   (B) No
   (C) Don't know

11. In order to realize sustainable development (e.g., to protect natural resources as much as possible), do you think it is necessary to introduce independent third-party organizations (e.g., the following forms) from the start of bidding? Please use a scale of 1 (strongly disagree) to 4 (strongly agree) to indicate your approval.

	Strongly disagree	Strongly agree	Do not know
(1) Specialized agency			
(2) Audit company			
(3) Guild			
(4) Non-governmental organizations (NGO) of civil society			

12.   As far as you know, is there a special supervision department for public-private joint projects in China or your industry and is responsible to the government?

   (A) Yes
   (B) No

13. Public-private cooperation can bring innovation and technology transfer to the region. Do you agree with this view?

   Answer:

14. In terms of infrastructure and public services, compared with the traditional government operation mode, public-private partnerships can create more job opportunities. Do you agree with this view?

Answer:

15. In China or in a specific industry, is it sufficient to provide public-private partnership training for local people with public-private partnership professionals?

    (A) Yes
    (B) No
    (C) Don't know

16. Are trade unions, civil society organizations and other non-governmental organizations involved in the development of public-private partnership projects?

    (A) Yes
    (B) No
    (C) Don't know

17. The government and the private sector share responsibility for security, but it is more difficult to achieve security. Do you agree with this view?

    Answer:

18. Does the original PPP contract include security measures?

    (A) Yes
    (B) No
    (C) Don't know

19. Here are some factors that can explain why your country lacks purchasing power parity. Could you please evaluate the importance of these factors? Please use a scale of 1 (unimportant) to 4 (very important).

    • Lack of knowledge and information ( ).
    • Overall lack of attractiveness to private investors ( ).
    • In the absence of a profitable market, it is not attractive to private investors ( ).
    • The lack of private investors is the result of very distant and low return on investment ( ).
    • Lack of legal system and procedures to stimulate investor confidence ( ).
    • There is a lack of government regulations that can prevent bad public-private partnership agreements, mismanagement, or corruption ( ).

20. Is the private sector prepared to share the risks of providing public services?

    (A) No preparation
    (B) Is not ready yet.
    (C) Is basically ready
    (D) Is completely ready
    (E) Don't know

21. Is the national legal system clear enough to support the development of public-private partnerships?

    (A) Yes
    (B) No
    (C) Don't know

22. Is China's existing system (policies, procedures, and services) developed enough to support the development of the PPP project?

    (A) Yes
    (B) No
    (C) Don't know

23. Your main job is to:

    (A) Government departments, agencies, or other public sector
    (B) Private (for-profit) business sector
    (C) Non-governmental (nonprofit) organization
    (D) Non-governmental social, religious, or cultural association

24. Your education level:

    (A) Compulsory education
    (B) Diploma in vocational and technical (engineering)
    (C) Business school diploma
    (D) University diploma
    (E) Graduate student

Bibliography

Municipal Public Utilities Research Center (2004) International experiences on the supervision for the municipal and public infrastructures [R]

Shanghai International Group Co., Ltd (2008) Shanghai urban construction investment and development corporation industry public-private partnership [R], vol 6

Yifu L, Chaoyang X (2005) Policy burden and soft budget constraint [R]. Working Paper of China Economic Research Center of Peking University

Yifu L, Guofu T (1999) Endogenous ability, policy burden, responsibility attribution and soft budget constraint [R]. Discussion papers series of china economic research center of Peking university. 06

© Science Press & Springer Nature Singapore Pte Ltd. 2020
Z. Meng et al., *Public Private Partnership for Desertification Control in Inner Mongolia*, https://doi.org/10.1007/978-981-13-7499-9